T0303343

How, When, and Whether to Employ Non-Lethal Weapons in Diverse Contexts

SCOTT SAVITZ, KRISTA ROMITA GROCHOLSKI, MONIKA COOPER,
NANCY HUERTA, KEYTIN PALMER, ISABELLE WINSTON

Prepared for the Joint Intermediate Force Capabilities Office
Approved for public release; distribution is unlimited

NATIONAL DEFENSE RESEARCH INSTITUTE

For more information on this publication, visit **www.rand.org/t/RRA2721-1**.

About RAND

The RAND Corporation is a research organization that develops solutions to public policy challenges to help make communities throughout the world safer and more secure, healthier and more prosperous. RAND is nonprofit, nonpartisan, and committed to the public interest. To learn more about RAND, visit www.rand.org.

Research Integrity

Our mission to help improve policy and decisionmaking through research and analysis is enabled through our core values of quality and objectivity and our unwavering commitment to the highest level of integrity and ethical behavior. To help ensure our research and analysis are rigorous, objective, and nonpartisan, we subject our research publications to a robust and exacting quality-assurance process; avoid both the appearance and reality of financial and other conflicts of interest through staff training, project screening, and a policy of mandatory disclosure; and pursue transparency in our research engagements through our commitment to the open publication of our research findings and recommendations, disclosure of the source of funding of published research, and policies to ensure intellectual independence. For more information, visit www.rand.org/about/principles.

RAND's publications do not necessarily reflect the opinions of its research clients and sponsors.

About This Report

Non-lethal weapons (NLWs) can be used to influence individuals' behavior and/or to suppress, degrade, or impair the performance of personnel and materiel. Prior RAND reports characterized how employment of NLWs contributes to the strategic goals of the U.S. Department of Defense (DoD), how NLWs' impact can be measured, and how they can be effectively integrated into wargames. However, there are few data points from either real-world or wargaming contexts (beyond military policing) regarding how NLWs could or should be employed in various contexts. To address this gap, the Joint Intermediate Force Capabilities Office (JIFCO) asked RAND to conduct a vignette-centered analysis of using NLWs in a range of operational and strategic contexts, drawing on research related to psychology and group dynamics, which is described in this report.

The research reported here was completed in December 2023, after which it was reviewed by both internal and external experts, as well as by JIFCO. It underwent security review with the sponsor and the Defense Office of Prepublication and Security Review before public release.

RAND National Security Research Division

This research was sponsored by JIFCO and conducted within the Navy and Marine Forces Program of the RAND National Security Research Division (NSRD), which operates the National Defense Research Institute (NDRI), a federally funded research and development center sponsored by the Office of the Secretary of Defense, the Joint Staff, the Unified Combatant Commands, the Navy, the Marine Corps, the defense agencies, and the defense intelligence enterprise.

For more information on the RAND Navy and Marine Forces Program, see www.rand.org/nsrd/nmf or contact the director (contact information is provided on the webpage).

Acknowledgments

The authors would like to acknowledge JIFCO, which funded this study. We received thoughtful guidance and vital information from many personnel at JIFCO, including Susan LeVine, COL Wendell Leimbach, Timothy Fox, Robert O'Day, Ryan Wolfgram, John Nelson, and Eric Duckworth. We would like to particularly acknowledge Shannon Foley, Douglas Peters, and Wesley Burgei from the JIFCO Human Effects division for participating in multiple interviews and discussions regarding their expertise on the psychological and behavioral impacts of NLWs. We would also like to thank the RAND program directors who oversaw this work, Paul DeLuca and Brendan Toland, for their continued support and comments. Finally, we want to thank our reviewers, Angela Putney and Peter Dobias, for providing thoughtful feedback that helped us to improve the report. We also appreciate the informal feedback we received from other colleagues, including Nathan Beauchamp-Mustafaga, Jeffrey Hornung, Bradley Martin, and J.D. Williams, which further strengthened this work.

Summary

Issue

Non-lethal weapons (NLWs) can be used to influence individuals' behavior and/or to suppress, degrade, or impair the performance of personnel and materiel. Examples of NLWs include acoustic hailers that warn, laser dazzlers that create eye-safe glare, and electronic systems that remotely shut down vehicles or vessels. As RAND prior research indicates, using NLWs can contribute to the strategic goals of the U.S. Department of Defense.[1] However, NLWs have rarely been used beyond military policing contexts, and there has been limited exploration by the Department of Defense on how to use NLWs effectively in various situations. To address this shortfall, the Joint Intermediate Force Capabilities Office (JIFCO) asked RAND to build on its previous work by conducting a vignette-centered analysis of NLW usage in a range of operational and strategic contexts. Given that many NLWs are intended to produce changes in individual and group behaviors, much of this analysis draws on research regarding the psychology of risk perception, decisionmaking, and group dynamics. This report describes the results of our analysis, which addresses how NLWs can be used effectively in different types of operational environments.

Approach

We gathered information about relevant aspects of psychology, group dynamics, and the human effects of NLWs from both documents and interviews. By analyzing this information and prior work, we developed diverse vignettes to explore how different factors might influence the impact and effectiveness of various modalities of employing NLWs. We then conducted a series of internal workshops to analyze how NLWs could be employed in the vignettes, documenting not only the opportunities to use NLWs effectively but also the various constraints and possible desirable future capabilities. Finally, we extracted our overarching findings from analyzing the vignettes.

---

[1] Krista Romita Grocholski, Scott Savitz, Jonathan P. Wong, Sydney Litterer, Raza Khan, and Monika Cooper, *How to Effectively Assess the Impact of Non-Lethal Weapons as Intermediate Force Capabilities*, RAND Corporation, RR-A654-1, 2022; Krista Romita Grocholski, Scott Savitz, Nancy Huerta, Alyson Youngblood, Jonathan P. Wong, Sydney Litterer, Raza Khan, and Monika Cooper, *Logic Model for Non-Lethal Weapons in the U.S. Department of Defense*, RAND Corporation, DV-A1544-1, 2023.

Key Findings

- NLWs can be effective in a range of operational contexts, spanning gray-zone confrontations, civilian encounters, and even combat, as well as situations in which civilians are present during gray-zone or combat operations.
- Two NLWs, acoustic hailing devices (AHDs) and laser dazzlers, were useful across virtually all of these diverse vignettes.
- Many current NLW systems have limited abilities to affect either large ships or individuals at appreciable distances who are not within a direct line of sight.
- Effective NLW tactics and concepts of employment (CONEMPS) need to incorporate the results of research on human effects, psychology, and group dynamics. Shaping perceptions, taking advantage of people's mental shortcuts in making decisions, and employing well-honed approaches to persuasion can help to influence behavior.
- NLWs can be particularly useful in confined, crowded, or complex environments. However, very close ranges can cause some NLWs (e.g., AHDs) to affect U.S. forces in addition to desired populations.
- Although countermeasures exist against some NLWs, they offset only parts of the effects, and individuals who employ countermeasures limit their own ability to perceive sensory information, degrading the countermeasures' capabilities.
- It would be highly desirable to mount AHDs and laser dazzlers aboard uncrewed aerial vehicles (UAVs).
- UAVs themselves and various onboard NLW capabilities were useful across multiple contexts, particularly in dealing with confined physical environments and large groups of people.
- Additional future capabilities that were desirable in our analysis included capabilities that could be employed against large ships, automated systems that could point multiple laser dazzlers at several moving or stationary targets, and a more portable, battery-powered Active Denial System.

Recommendations

Our overarching recommendations are for JIFCO to

- **incorporate psychological, group-dynamic, and human-effect insights** into plans, CONEMPS, and tactics for using NLWs that affect human beings and their behavior, including consideration of adversary perspectives and approaches to communicating that NLWs' effects are reversible[2]

---

[2] The effects of antimateriel NLWs, as well as physiologically incapacitating NLWs such as Tasers, are more readily understood without drawing on psychological and group-dynamic research. However, it can be useful to analyze the psychological or group-dynamic effects of incapacitating either materiel or other individuals.

- **prioritize the highly versatile AHDs and laser dazzlers** for service acquisition, training, and fielding, ensuring that both are highly integrated into operations
- **pursue such desirable future capabilities as UAV-based NLWs, Maritime Vessel Stopping Occlusion Technologies (MVSOT)–like capabilities against large ships, automated aiming of laser dazzlers, and a smaller, lighter Active Denial System**
- **ensure legal review** of NLW CONEMPS and tactics because, in some cases, employment of NLWs might be impeded by legal strictures
- **get more data** to support service-specific development of tactics and CONEMPS for NLW usage via modeling, wargaming, live exercises, and use in real-world operations, as appropriate.[3]

---

[3] Modeling that incorporates psychological and group-dynamic principles could be particularly valuable, such as the Workbench for refining Rules of Engagement against Crowd Hostiles (WRENCH) model developed by the Naval Postgraduate School and JIFCO. See Susan Aros, Anne Marie Baylouny, Deborah E. Gibbons and Mary McDonald, "Toward Better Management of Potentially Hostile Crowds," paper presented at the 2021 Winter Simulation Conference, 2021; and Susan Aros and Mary McDonald, "Simulating Civil Security Activities in Stability Operations," paper presented at the Interservice/Industry Training, Simulation, and Education Conference, November 2023.

Contents

Figures and Table

Figures

Table

CHAPTER 1

Introduction

Background

Non-lethal weapons (NLWs) can be used to influence individuals' behavior or to temporarily incapacitate either people or equipment, and NLWs employ a wide variety of effects to achieve these aims. For example, an acoustic hailing device (AHD) can be used to communicate or generate irritating sounds, an eye-safe ocular interrupter (OI) laser dazzler creates distracting glare, an Active Denial System (ADS) emits millimeter-wave energy to create a temporary heating sensation, and various electronic and mechanical systems can halt vehicles or vessels. NLWs are a subset of intermediate force capabilities (IFCs), which include other capabilities that might not have lethal effects, such as information operations, electronic warfare, and cyberwarfare.

Prior RAND reports have characterized how employment of NLWs contributes to the strategic goals of the U.S. Department of Defense (DoD).[1] Those reports also describe ways to measure the impact of NLWs, as well as how they can be effectively incorporated into wargames. Those reports also reveal a number of reasons why DoD rarely employs NLWs. One of these is widespread unfamiliarity with these systems, which contributes to a reluctance to bring them to forward positions or to use them, continuing that unfamiliarity in a vicious cycle. The result is that there are relatively few well-documented instances of NLWs being employed in real-world contexts beyond the scope of military policing. The limited amount of real-world data is exacerbated by the relatively few instances in which NLWs have been incorporated into DoD wargames thus far, meaning that there has been little exploration of how these systems can be used. The North Atlantic Treaty Organization (NATO) has

---

[1] Krista Romita Grocholski, Scott Savitz, Jonathan P. Wong, Sydney Litterer, Raza Khan, and Monika Cooper, *How to Effectively Assess the Impact of Non-Lethal Weapons as Intermediate Force Capabilities*, RAND Corporation, RR-A654-1, 2022; Krista Romita Grocholski, Scott Savitz, Sydney Litterer, Monika Cooper, Clay McKinney, and Andrew Ziebell, *Assessing the Impact of Diverse Intermediate Force Capabilities and Integrating Them into Wargames for the U.S. Department of Defense and NATO*, RAND Corporation, RR-A1544-1, 2023.

For a visualization of the logic model that links NLW activities to direct outputs, higher-level outcomes, and ultimate strategic goals, see Krista Romita Grocholski, Scott Savitz, Nancy Huerta, Alyson Youngblood, Jonathan P. Wong, Sydney Litterer, Raza Khan, and Monika Cooper, *Logic Model for Non-Lethal Weapons in the U.S. Department of Defense*, RAND Corporation, DV-A1544-1, 2023.

done more work in this regard in recent decades; its Systems Analysis and Studies program has completed a series of studies and recent wargames on NLWs.[2] These efforts defined and characterized NLWs and other IFCs, identified NLW requirements, validated NLW contributions to mission success, assessed capability deficiencies, and identified areas for investment in research and development.[3] However, there is a paucity of information and insight regarding how NLWs could or should be employed by DoD in various contexts, given that the United States sometimes has different needs from its NATO partners.

Purpose of This Report

To address this shortfall, the Joint Intermediate Force Capabilities Office (JIFCO) asked RAND to build on its previous work by conducting a vignette-centered analysis of NLW usage in a range of operational and strategic contexts that are relevant to DoD. Given that many NLWs are intended to produce changes in individual and group behaviors, much of this analysis draws on research regarding the psychology of risk perception, decisionmaking, and group dynamics. This report describes the results of our analysis.

Methodology

We gathered information regarding the relevant aspects of psychology, group dynamics, and the human effects of NLWs from documents and interviews. We used this information and the results of our previous research on NLWs at RAND to conduct a vignette-centered analysis.[4] For this analysis, we developed a series of diverse vignettes that cover different operational environments, operational scenarios, and other psychologically relevant factors to explore how those factors might influence the impact and effectiveness of various modalities of employing NLWs. We conducted a series of internal workshops to analyze how NLWs could be employed in the vignettes and documented opportunities to use NLWs effectively, as well as various constraints and potential future capabilities that might be desirable. The participants in these workshops included individuals at RAND who have expertise in NLWs, emerging technologies, psychology, group dynamics, and international relations. The vignettes and proposed approaches to NLW employment were also reviewed by RAND

---

[2] Peter Dobias and Kyle Christensen, "Intermediate Force Capabilities: Countering Adversaries Across the Competition Continuum," *Journal of Advanced Military Studies*, Vol. 14, No. 1, November 2023. Some of the recent wargames illustrated the impact and utility of NLWs and IFCs in multiple contexts, such as force protection, maritime operations, and contested non-combatant evacuations for NATO forces.

[3] Dobias and Christensen, 2023.

[4] For these prior works, see Romita Grocholski et al., 2022; and Romita Grocholski, Savitz, Litterer, et al., 2023.

colleagues with additional expertise in the gray zone, maritime operations, NLWs, and psychology, as well as by experts from outside RAND.

Key Definitions and Overview of NLWs

Before delving into the substance of this report, we briefly define and explain NLWs. Then, we briefly describe the gray zone, a key context in which NLWs are used.

Defining *NLWs*

According to Department of Defense Directive 3000.03, *NLWs* are defined as follows:[5]

> Weapons, devices, and munitions that are explicitly designed and primarily employed to incapacitate targeted personnel or materiel immediately, while minimizing fatalities, permanent injury to personnel, and undesired damage to property in the target area or environment. NLWs are intended to have reversible effects on personnel and materiel.

In a prior RAND report, we proposed a slightly different definition for future consideration by DoD.[6] That definition was intended to encompass all IFCs, including NLWs, information operations, electronic warfare, and cyberwarfare. Below, we have revised the definition slightly to be more specific to NLWs. In comparison with the DoD definition above, this version highlights the relevance of these systems across the spectrum of conflict to include peacetime, as well as their effects on behavior, which might not solely stem from incapacitation:

> Systems and capabilities that can be used in all phases of conflict to suppress, degrade, or impair the performance of personnel and material and/or to influence individuals' behavior by producing predictable, immediate, reversible effects that are intended to minimize harm or permanent damage.

Regardless of which definition is used, NLWs represent a highly diverse set of capabilities that use various modalities to affect human behavior or contribute to temporary incapacitation of humans or equipment. A few of these NLWs were mentioned previously, but we explain them at greater length below, recapitulating from prior RAND reports:[7]

- **Acoustic systems.** These systems project sound for various purposes. They can be used to communicate and try to change behavior—e.g., to try to get members of a large group

---

[5] Department of Defense Directive 3000.03E, *DoD Executive Agent for Non-Lethal Weapons (NLW) and NLW Policy*, U.S. Department of Defense, incorporating change 2, August 31, 2018, p. 12.

[6] Romita Grocholski et al., 2022.

[7] Romita Grocholski et al., 2022; Romita Grocholski, Savitz, Litterer, et al., 2023.

to back away or to get a ship's captain to change course to avoid possibly ramming a U.S. Navy ship. Acoustic systems can also be used to emit distracting, irritating, or simply loud sounds that try to shape the behavior of other parties. Existing acoustic systems include the AHD and the long-range acoustic device (LRAD).[8] These devices are small enough to be easily transported by personnel or vehicles and can be used either to amplify speech or to emit prerecorded sounds or messages. The experimental Laser-Induced Plasma Effects (LIPE) is intended to use lasers to create sounds that emerge from a specified location at a distance.[9]

- **Laser dazzlers.** These weapons create an intense glare that distracts and limits perception; the effects are a more intense version of having the sun in a person's eyes. Two laser dazzler systems, the OI and long-range ocular interrupter (LROI), have been tested to ensure that they do not have any permanent effects; therefore, both are compliant with the Protocol on Blinding Laser Weapons.[10] In addition to degrading the capabilities of a possible foe, these devices can be used to warn and to differentiate intent, depending on whether dazzled individuals back away or continue to approach. The OI can easily be held in a person's palm, whereas the LROI requires two hands.[11]
- **Integrated-effects systems.** These systems combine acoustic and laser effects, as well as nonlaser spotlights that can be used to illuminate and warn. Two current systems are the Hailing Acoustic Laser and Light Tactical System (HALLTS) and the developing Escalation of Force Common Remotely Operated Weapons Station (EoF CROWS).[12]
- **Flash-bang grenades.** These weapons generate intense sound and light, which can distract and temporarily incapacitate.[13]
- **Blunt-impact munitions.** These munitions are designed to inflict only limited injuries when they hit people. Examples include rubber bullets, grenades that disperse rubber pellets, paintballs, and "beanbag" rounds.[14]

---

[8] JIFCO, DoD, Non-Lethal Weapons Program Office, "Acoustic Hailing Devices Fact Sheet," November 16, 2018a.

[9] Patrick Tucker, "The US Military Is Making Lasers That Create Voices out of Thin Air," Defense One, March 20, 2018.

[10] Protocol on Blinding Laser Weapons (Protocol IV to the 1980 Convention), signed October 13, 1995.

[11] JIFCO, DoD, Non-Lethal Weapons Program, "Non-Lethal Optical Distractors Fact Sheet," May 2016; Office of General Counsel, U.S. Department of Defense, *Department of Defense Law of War Manual*, updated May 2016, pp. 411–412.

[12] DoD, Physical Security Enterprise and Analysis Group, "Hailing Acoustic Laser & Light Tactical System (HALLTS)," webpage, undated; JIFCO, DoD, Non-Lethal Weapons Program, *DoD Non-Lethal Capabilities: Enhancing Readiness for Crisis Response: Annual Review*, 2015.

[13] JIFCO, DoD, Non-Lethal Weapons Program, 2015.

[14] JIFCO, DoD, Non-Lethal Weapons Program, 2015.

- **Electromuscular incapacitation systems.** Tasers and stun guns generate electrical currents that cause muscles to contract, which temporarily incapacitates the target.[15]
- **Non-lethal chemical irritants.** These items, such as pepper spray and tear gas, are typically reserved for law-enforcement situations in which it is desirable to temporarily incapacitate people. These irritants are sometimes termed *riot-control agents* and are largely prohibited from being used by the military by both the Chemical Weapons Convention—which forbids their use as a method of warfare—and Executive Order 11850.[16] However, the executive order also permits the use of such agents with presidential approval for cases in which civilians are being used to shield attackers and civilian lives can be saved, to protect movements in non-combat areas from terrorist or other attacks, in rescue missions in isolated areas, or against rioting prisoners of war.[17] However, the use of such weapons in an ambiguous context would likely be met with widespread hostility, given the popular aversion to chemical warfare.[18]
- **Millimeter-wave systems.** The ADS uses a beam of millimeter-wave energy to induce a heating sensation that dissipates as soon as a person steps out of the beam or the beam is turned off. A more portable version of the system that is being developed is called ADS Solid State.[19]
- **Microwave systems.** The Radio Frequency Vehicle Stopper (RFVS) and the Pre-Emplaced Vehicle Stopper (PEVS) emit microwave beams to temporarily disable ground vehicles' electronics, whereas the Vessel Incapacitating Power Effect Radiation (VIPER) does the same for small maritime vessels.[20]

---

[15] JIFCO, DoD, Non-Lethal Weapons Program, "Human Electro-Muscular Incapacitation FAQs," webpage, undated-b.

[16] Convention on the Prohibition of the Development, Production, Stockpiling and Use of Chemical Weapons and on Their Destruction, signed at Geneva, Switzerland, September 3, 1992; Executive Order 11850, "Renunciation of Certain Uses in War of Chemical Herbicides and Riot Control Agents," Executive Office of the President, April 8, 1975.

[17] JIFCO, DoD, Non-Lethal Weapons Program, "Oleoresin Capsicum Dispensers," webpage, undated-c; JIFCO, DoD, Non-Lethal Weapons Program, "Variable Kinetic System (VKS) Non-Lethal Launcher System," webpage, undated-d.

[18] The U.S. government has reacted strongly against the use of lethal chemical weapons in the past: Secretary of State John Kerry declared that Syria's use of them in 2013 constituted a "global red line" (Voice of America, "Kerry: Syrian Gas Attack Crossed 'Global Red Line,'" September 8, 2013). Although the Syrian regime had slaughtered much larger numbers of people using other types of weapons, killing and injuring people with chlorine was perceived as a potential casus belli.

[19] JIFCO, DoD, Non-Lethal Weapons Program, "Active Denial System FAQs," webpage, undated-a; Susan LeVine, *The Active Denial System: A Revolutionary, Non-Lethal Weapon for Today's Battlefield*, National Defense University Center for Technology and National Security Policy, June 2009.

[20] Jamal Beck, "New Vehicle Stopper Trials Underway at Tinker Air Force Base," press release, Joint Non-Lethal Weapons Directorate, August 15, 2018; JIFCO, DoD, Non-Lethal Weapons Program, "Vessel-Stopping Prototype," November 16, 2018d; JIFCO, DoD, Non-Lethal Weapons Program, "Radio Frequency Vehicle Stopper," November 16, 2018b.

- **Mechanical vehicle- or vessel-stopping technologies.** One system, the Single Net Solution with Remote Deployment Device (SNS–RDD), pierces vehicle tires with a spiked net; simple "spike strips" can also be placed in a given pathway to achieve the same effect. JIFCO is also pursuing Maritime Vessel Stopping Occlusion Technologies (MVSOT) that either coat propellers with a sticky substance or entangle them in long, durable cables.[21]

A key aspect of NLWs that project sound or electromagnetic energy, such as acoustic systems, laser dazzlers, ADSs, RFVSs, and VIPERs, is the degree of directionality of their effects. Sounds generated by AHDs have directionality—some areas are subject to louder sound than others—but generally have relatively wide-angle coverage, and some of their effects can be felt from any angle. By contrast, the NLWs that emit electromagnetic beams (i.e., laser dazzlers, ADSs, RFVSs, and VIPERs) are highly directional and intended to only affect a single person or vehicle. Sounds can affect a large group simultaneously, whereas electromagnetic beams need to be targeted against specific individuals. For the antipersonnel laser dazzler and ADS, this targeting likely entails focusing on those who are in command or other leadership roles, or focusing on those who are agitating, attacking, intoxicated, or refusing to depart.

A second way to characterize NLWs is in terms of their susceptibility to countermeasures. Sound is hard to screen out: Although people can put in earplugs or noise-canceling headphones (thereby also restricting their access to other auditory information), they can still feel vibrations. Special glasses or simply closing one's eyes can be used to counter the glare from laser dazzlers, although the latter comes at the price of losing access to other visual information. No comparable countermeasures—other than being surrounded by shielding material—are available for ADS.

Overall, the fact that NLWs are designed to limit the risk of fatalities or permanent injuries can help to make their employment less escalatory than the use of lethal weapons, as highlighted by both NATO and RAND studies.[22] Studies of NATO and RAND analyses describe how NLWs can also help to protect forces while limiting the risk of harm to civilians.[23]

The extent or types of escalation that might occur are inevitably influenced by the perceptions of those experiencing NLW effects. If targeted individuals misperceive NLWs as lethal or as inflicting permanent injury, or if they simply react without a clear understanding of what they are experiencing, they might undertake escalatory lethal actions. However, in the vignettes that follow, we assume that the targeted side perceives that the NLWs do not induce

---

[21] Army Nonlethal Scalable Effects Center, "Army Nonlethal Weapons Advanced Planning Briefing to Industry," U.S. Army, September 21, 2017; Katherine Mapp, "Promising New Tool Protects Ships, Sailors," Naval Sea Systems Command, November 21, 2019; Nathan Gain, "US Navy Lab Investigates Innovative Non-Lethal Boat Stopping Technology," Naval News, November 25, 2019; JIFCO, DoD, Non-Lethal Weapons Program, "Single Net Solution with Remote Deployment Device," November 16, 2018c.

[22] Dobias and Christensen, 2023; Romita Grocholski, Savitz, Litterer, et al., 2023.

[23] Dobias and Christensen, 2023; Romita Grocholski, Savitz, Litterer, et al., 2023; Romita Grocholski et al., 2022.

fatalities and that the temporary nature of their effects quickly becomes apparent, a point that is reinforced by communications from U.S. forces. In this context, it makes sense that an adversary would likely perceive the use of NLWs as less of an escalatory step than the use of lethal force. However, analysis of the perspectives of adversaries based on intelligence and on what they are able to perceive at any given moment needs to be taken into account.

Defining the *Gray Zone*

A key type of operational environment in which NLWs can be useful is the gray zone, a space between wholly peaceful interactions and full-scale war. Much of our work in this report involves situations in the gray zone, so it is important to understand what a gray-zone scenario entails from the outset. Although there is no universally shared definition of the *gray zone*, in keeping with its ambiguity, we offer three closely aligned definitions developed by our RAND colleagues. The first defines *gray-zone aggression*, the second defines *gray-zone tactics*, and the third defines the *gray zone* itself.

> We define [*gray-zone aggression*] as an integrated campaign to achieve political objectives while remaining below the threshold of outright warfare. Typically, such campaigns involve the gradual application of instruments of power to achieve incremental progress without triggering a decisive military response.[24]

> [*Gray-zone tactics* are] coercive . . . government geopolitical, economic, military, and cyber and information activities beyond regular diplomatic and economic activities and below the use of kinetic military force.[25]

> The gray zone is an operational space between peace and war, involving coercive actions to change the status quo below a threshold that, in most cases, would prompt a conventional military response, often by blurring the line between military and nonmilitary actions and the attribution for events.[26]

Regardless of which definition is used, the central feature is clear—a gray-zone situation involves military actions that change the situation but do not rise to the level of full-fledged war. In this report, we explore gray-zone situations in which military forces are confronting

---

[24] Michael J. Mazarr, Joe Cheravitch, Jeffrey W. Hornung, and Stephanie Pezard, *What Deters and Why: Applying a Framework to Assess Deterrence of Gray Zone Aggression*, RAND Corporation, RR-3142-A, 2021, p. 1.

[25] Bonny Lin, Cristina L. Garafola, Bruce McClintock, Jonah Blank, Jeffrey W. Hornung, Karen Schwindt, Jennifer D. P. Moroney, Paul Orner, Dennis Borrman, Sarah W. Denton, and Jason Chambers, *Competition in the Gray Zone: Countering China's Coercion Against U.S. Allies and Partners in the Indo-Pacific*, RAND Corporation, RR-A594-1, 2022, p. 2.

[26] Lyle J. Morris, Michael J. Mazarr, Jeffrey W. Hornung, Stephanie Pezard, Anika Binnendijk, and Marta Kepe, *Gaining Competitive Advantage in the Gray Zone: Response Options for Coercive Aggression Below the Threshold of Major War*, RAND Corporation, RR-2942-OSD, 2019, p. 8.

one another in ways that could lead to danger and escalation and in which U.S. forces could potentially use NLWs to help reduce risks and prevent rivals from achieving their goals.

Structure of This Report

Chapter 2 explores insights from psychological and group-dynamic research to highlight how NLWs can affect different types of groups as a result of the context and how NLWs are used. Chapter 3 provides some background on how activities employing NLWs contribute to higher-level goals and then provides an overview of the vignettes discussed in Chapters 4–6, which describe our analysis of a series of vignettes in which NLWs are used in different physical and operational environments. Chapter 4 focuses on vignettes involving gray-zone encounters, Chapter 5 covers vignettes involving combat, and Chapter 6 includes vignettes involving large groups of civilians. Finally, Chapter 7 provides a set of overarching findings and recommendations.

CHAPTER 2

Psychological and Group-Dynamic Insights

Although NLWs can be used to temporarily impair or incapacitate, much of their purpose is to change individual and/or group behaviors. Such groups could be assembled civilians or another nation's military units, including in the gray zone below the threshold for full-scale conflict or even in combat. Changing the behaviors of individuals or groups ultimately depends on some combination of communication and persuasion—i.e., conveying to someone that it is better for them to take a particular course of action. Changing people's behavior requires influencing their judgment regarding the desirability of courses of action and leading them to make decisions that are favorable. For example, changing people's behaviors might include getting a group of angry civilians to move out of the way of a convoy. Alternatively, in a gray-zone situation, changing an individual's behavior might entail getting the captain of a rapidly approaching warship to change course to avert a collision. It can even involve combat: In some situations, even while using lethal weapons against combatants, it might be valuable to use NLWs to influence combatants and/or civilians.

In the context of NLWs, this chapter describes some findings from psychology and group dynamics. That research can help to better understand factors that can influence the effectiveness of NLWs that are used to influence behavior. We apply our insights from this research in the vignette analysis that is described in Chapters 4–6. We begin by discussing the psychology of judgment and decisionmaking, then discuss the sensory aspects of response, and close by discussing how both affect group dynamics.

The Psychology of Judgment and Decisionmaking

To effectively change behavior using NLWs, it is critical to understand the basis for judgment and decisionmaking, which draws on extensive research over the last half century. One of the central findings is the extent to which human judgments and decisions are not grounded in deeply rational analysis.[1] Given limited time, numerous stressors, incomplete information,

---

[1] Amos Tversky, Daniel Kahneman, and their collaborators wrote an array of books and articles about these topics. Two collections of their work include Daniel Kahneman, Paul Slovic, and Amos Tversky, eds., *Judgment Under Uncertainty: Heuristics and Biases*, Cambridge University Press, 1982; and Amos Tversky and Daniel Kahneman, *Choices, Values, and Frames*, Cambridge University Press, 2000.

and various emotional cues, human beings often make decisions based on rapid intuition—termed *System 1* thinking—as opposed to more-deliberative assessments, which compose *System 2* thinking.[2] In the System 1 thinking that tends to dominate most human decision-making, people use mental shortcuts called *heuristics* to make quick judgments and decisions. Leading experts in the field—Daniel Kahneman, the late Amos Tversky, and many of their collaborators—have identified a number of such heuristics, some of which complement or overlap with one another. A few of these heuristics include the following:

- **Framing.** The way in which a question is framed can radically affect decisions. For example, Tversky and Kahneman conducted experiments in which people could choose between two options that would affect 600 people in imminent danger of death. The first option would consistently save 200 people, while the second option had a one-third probability of saving all 600 and a two-thirds probability of saving none. Fully 72 percent of participants preferred to save the 200, while the rest chose the second option. Other participants were provided with the exact same problem but posed in terms of the number of people dying rather than saved: The first option would result in 400 deaths, while the second option had a one-third probability of no deaths and a two-thirds probability of 600 deaths. Given that phrasing, only 22 percent of participants preferred the first option, and all the others chose the second. In other words, half of the population's choices depended on the way in which the problem was phrased.[3]
- **Availability.** People tend to perceive the likelihood of events as correlated with their ability to envision them (i.e., the mental "availability" of those possibilities).[4]
- **Anchoring.** People's assessments are often heavily biased by the initial data points.[5]
- **Representativeness.** People often perceive that any small sample of data is indicative of the larger whole, even though small samples can be highly biased.[6]
- **Overconfidence.** People often overestimate the accuracy of their assessments, including within their areas of expertise, such as medical diagnoses.[7] People are also prone to overstate the extent of their knowledge of areas beyond their expertise. The experimentally observed Dunning-Kruger effect (named for the researchers who identified it)

---

[2] Daniel Kahneman, *Thinking, Fast and Slow*, Farrar, Straus and Giroux, 2013.

[3] Amos Tversky and Daniel Kahneman, "The Framing of Decisions and the Psychology of Choice," *Science*, Vol. 211, No. 4481, January 1981.

[4] Amos Tversky and Daniel Kahneman, "Availability: A Heuristic for Judging Frequency and Probability," *Cognitive Psychology*, Vol. 5, No. 2, September 1973.

[5] Kahneman, Slovic, and Tversky, 1982.

[6] Kahneman, Slovic, and Tversky, 1982.

[7] Scott Plous, *The Psychology of Judgment and Decision Making*, McGraw-Hill, 1993; Kahneman, Slovic, and Tversky, 1982.

is that the less people know about a given subject, the more confident they are in their knowledge about it.[8]

From the standpoint of employing NLWs, the goal is to convey messages that work with these heuristics to help shape behaviors. For example, when faced with a group of people blocking a convoy's path, AHDs can be used to communicate in a way that frames choices: Staying here exposes you to risk, but going to a nearby aid station will enable you to get food and supplies. NLWs can also use these heuristics to shape behaviors during gray-zone encounters when another nation's forces are conducting operations below the threshold of war. For example, rapidly introducing multiple sensory stimuli at an early stage can intensely affect lasting perceptions and give the impression of large NLW capacities, which exploits the anchoring and representativeness heuristics. Doing so can help to shatter an opposing commander's prior assessments that U.S. forces will meekly back away, which shows that such assessments are overconfident both in terms of gauging relative resolve and in the opposing force's commander's perceptions of the probable accuracy of their prior assessments.

Other psychological insights can also be applied to using NLWs. In a leading book on changing perceptions and behaviors, *Influence: The Psychology of Persuasion*, Robert Cialdini identifies six broad approaches to influencing others:[9]

- **Reciprocity.** People want to see that others are treating them as they treat others.
- **Scarcity.** People want something in inverse proportion to its availability.
- **Authority.** People are more inclined to be influenced by those they perceive as authority figures.
- **Commitment and consistency.** People want to behave the same way over time.
- **Liking.** People are more influenced by those whom they like.
- **Consensus, or social proof.** People want to behave according to the norms that others follow.

Some of these approaches can be applied to NLW usage. For example, NLWs can be used against a large group of civilians in a way that demonstrates consistency: Only the people throwing projectiles will be subjected to glare-creating laser dazzlers. The types of auditory

---

[8] See Justin Kruger and David Dunning, "Unskilled and Unaware of It: How Difficulties in Recognizing One's Own Incompetence Lead to Inflated Self-Assessments," *Journal of Personality and Social Psychology*, Vol. 77, No. 6, 1999; and David Dunning, "The Dunning–Kruger Effect: On Being Ignorant of One's Own Ignorance," in James M. Olson and Mark P. Zanna, eds., *Advances in Experimental Social Psychology*, Elsevier Academic Press, Vol. 44, 2011.

There are also studies indicating that the extent of overconfidence can vary across cultures, at least in some contexts. For example, see Don A. Moore, Amelia S. Dev, and Ekaterina Y. Goncharova, "Overconfidence Across Cultures," *Collabra: Psychology*, Vol. 4, No. 1, October 2018; and Daniella Acker and Nigel W. Duck, "Cross-Cultural Overconfidence and Biased Self-Attribution," *Journal of Socio-Economics*, Vol. 37, No. 5, 2008. However, a detailed exploration of this topic would be beyond our scope.

[9] Robert B. Cialdini, *Influence: The Psychology of Persuasion*, 3rd ed., Quill, 1993.

messaging that AHDs project should sound both authoritative in the local language and respectful toward the audience. This messaging might not achieve mutual liking, but it can help to diminish hostility. Such messaging can also include potential rewards for desired behavior, as mentioned previously, which contributes to a degree of reciprocity. NLWs can also exploit social proof. If people are being told to disperse, inducing a few of them to do so can cause a cascading effect because others start to perceive departure as the norm rather than feel a social pressure to adhere to the prior norm of remaining.

Situation Awareness

The psychological term *situation awareness* (SA) refers to people's ability to perceive, understand, and respond to various circumstances in their environment.[10] Confusingly, DoD uses the term *situational awareness* with a different meaning, denoting a knowledge of the environment and forces within it.[11] In this report, we use the psychological definition because it is a useful framework for considering the impact of NLWs.

SA is broken down into three phases or levels—Level 1: *perceiving* information in one's environment based on sensory stimuli, Level 2: *comprehending* that information, and Level 3: accurately *projecting* future states of the environment—before people then make a decision about what actions to take.[12] These decisions and actions might be influenced by individuals' past experiences, knowledge, cultural norms, capabilities, emotional state, motivations, and goals.

SA Level 1: Perception

The first step in perceiving what is happening around us is sensation, which is the process of absorbing such information as light, sound, vibration, and pressure through our sensory receptors that is interpreted by the brain and perceived as images, sounds, smells, and tactile experiences.[13] NLWs can directly shape those perceptions in both positive and negative ways. Sounds emitted by AHDs can communicate, soothe, or irritate. Eye-safe laser dazzlers can distract and impair individuals while also effectively warning them to go away. The ADS induces a temporary tactile heating sensation.

---

[10] Mica R. Endsley, "Toward a Theory of Situation Awareness in Dynamic Systems," *Human Factors: The Journal of the Human Factors and Ergonomics Society*, Vol. 37, No. 1, March 1995.

[11] The DoD lexicon does not define *situational awareness*, although it does define it in specific contexts; see DoD, *DOD Dictionary of Military and Associated Terms*, Joint Chiefs of Staff, March 2017, p. 218, which defines *space situational awareness* as "[c]ognizance of the requisite current and predictive knowledge of the space environment and the operational environment upon which space operations depend."

[12] Endsley, 1995.

[13] Christopher D. Wickens, *Engineering Psychology and Human Performance*, 2nd ed., HarperCollins Publishers, 1992.

To perceive what is around us, we must not only sense it but also must attend, or pay attention, to it. This concept is demonstrated in a well-known psychology study in which participants were asked to view a video recording and count the number of times that people passed a basketball.[14] During the recording, a man in a gorilla suit walked directly through the frame, an event that most participants failed to perceive, because their attention was on the basketball. Simply because we lay our eyes on an object does not mean we actually perceive it; we must also pay attention to it. The act of attending to information might be something that happens as a result of conscious effort, or it might happen automatically.[15]

In this context, one potential use of NLWs is to overwhelm attentional resources by creating distractions across various senses. For example, when military forces are faced with adversaries in a gray-zone situation or combat, NLWs can create *cognitive overload* that overwhelms their adversaries' ability to perceive or process. This overloading of sensations can be focused on targeting the senses that their adversaries most need for particular tasks, if those are known, because distractions disproportionately degrade the efficiency of activities involving the senses that they affect.[16]

Various factors can affect how well individuals perceive NLW effects. Visual or auditory impairments, boredom, and fatigue may limit perception.[17] For instance, imagine an exhausted driver blithely driving past a red stoplight. Conversely, when people are in a high arousal state, such as anxiety, frustration, or anger, they are more likely to notice signals when they are not present (i.e., false positives).[18] Despite these risks, evoking such emotional states as anxiety might be an effective way to get members of a group to retreat or to force aggressors to desist.[19]

SA Level 2: Comprehension

After information is perceived, the sensations must be interpreted and comprehended to inform decisions and actions. Culture and language can also strongly influence this process; naturally, messages should be presented in the local dialect(s) and supplemented with visual signs, if possible. Using words in combination with universally recognized symbols or icons

---

[14] Daniel J. Simons and Christopher F. Chabris, "Gorillas in Our Midst: Sustained Inattentional Blindness for Dynamic Events," *Perception*, Vol. 28, No. 9, September 1999.

[15] Wickens, 1992.

[16] Basil Wahn and Peter König, "Is Attentional Resource Allocation Across Sensory Modalities Task-Dependent?" *Advances in Cognitive Psychology*, Vol. 13, No. 1, March 2017.

[17] Karl Halvor Teigen, "Yerkes-Dodson: A Law for All Seasons," *Theory & Psychology*, Vol. 4, No. 4, November 1994.

[18] Teigen, 1994.

[19] Mara Aruguete, "Non-Lethal Weapon Use in Crowds," in Lincoln University and the U.S. Army Research Laboratory, *Multi-Task Project to Provide Research Support for Human Research and Engineering Goals Identified by the Army*, 2009.

(e.g., arrows) can also speed up comprehension and increase understanding. Finally, well-recognized sounds, such as police sirens, are also nonverbal ways to communicate effectively with crowds and adversaries when there are language barriers.

SA Level 3: Projection

Projection refers to accurately predicting future states of an environment and how one's actions affect the future states. Projections are based on past experience or communication about future consequences of behaviors. There is a need to consistently communicate and demonstrate the consequences of both desirable and undesirable actions.

In a prior study, the author recommended maintaining credibility and strength by communicating consequences before administering them and consistently following through on any threats.[20] The people in the targeted group should perceive that the consequences of aggression are logical and that these consequences are being provoked by the aggressor. If the action is perceived to be unjust or unexpected, this might lead to others joining in the aggression in defense of the victim(s).[21] This recommendation aligns with some of Cialdini's principles: People want consistency, reciprocity, and fairness, and they will be angry if these principles are violated.

Decisions and Actions

After gaining SA, people decide what actions to take. By definition, decisionmaking involves selecting one option from a variety of alternatives when some information is available about the option you are choosing. However, the available information might be incomplete, complex, or distorted, and, furthermore, decisionmaking might be warped by the biases mentioned earlier, such as framing or anchoring.

---

[20] Elizabeth S. Mezzacappa, *Crowd Research and Military Interaction with Non-Lethal Weapons and Systems*, Target Behavioral Response Laboratory, JNLWD11-006, 2009.

[21] John Drury and Steve Reicher, "Collective Action and Psychological Change: The Emergence of New Social Identities," *British Journal of Social Psychology*, Vol. 39, No. 4, December 2000; John Drury and Steve Reicher, "The Intergroup Dynamics of Collective Empowerment: Substantiating the Social Identity Model of Crowd Behavior," *Group Processes & Intergroup Relations*, Vol. 2, No. 4, October 1999; Clifford Stott, Otto Adang, Andrew Livingstone, and Martina Schreiber, "Tackling Football Hooliganism: A Quantitative Study of Public Order, Policing and Crowd Psychology," *Psychology, Public Policy, and Law*, Vol. 14, No. 2, May 2008; Clifford Stott and Stephen Reicher, "Crowd Actions As Intergroup Process: Introducing the Police Perspective," *European Journal of Social Psychology*, Vol. 28, No. 4, December 1998.

Group Dynamics

Factors That Can Influence Group Behavior

There are factors that have been identified as correlating with a propensity for conflict escalation between personnel and large groups, which are drawn from interviews and focus groups with military and domestic law-enforcement personnel. Such factors as being predominantly male, relatively youthful, or intoxicated, or having evidence of competing factions, high noise levels, hostile body language, and fighting within the group, all suggest that a group is likely to be confrontational.[22] None of these factors means that there will be aggression, but they are more common in situations that escalate than in those that remain peaceful.

Another powerful factor influencing the behavior of large groups is mutual observation by individuals. The departure of a few individuals can inspire others to do the same, as we previously mentioned in connection with Cialdini's principle of social proof. A thought experiment also reveals how small changes in conditions can have a large impact on group behaviors. Mark Granovetter proposed a simple model of 100 people, in which one person will riot for no reason at all, a second person will riot if they see one other person rioting, a third person will riot if they see two people rioting, and so on until the 100th person will riot only if they see 99 people rioting.[23] Once the first person starts a riot, the second joins it, and all 100 eventually join. However, if we remove the second person from this model, only the first person riots, and the others do not.

Even though this is an extremely simplistic example, which Granovetter freely acknowledged, it points to the way in which people with different thresholds for particular behaviors can result in radically different outcomes from only minor shifts. For instance, if we hypothesize that members of one subgroup are highly receptive to messaging telling them to disperse, whereas members of a second subgroup will disperse only if they perceive others doing so and if they perceive that the most-aggressive individuals are being non-lethally targeted with laser dazzlers, it is possible that group behavior may change based on limited NLW effects.

Unsurprisingly, music can play a role in influencing the behavior of groups. Projecting soothing music via AHDs might be a way of defusing confrontations. Conversely, music can be an inducement to disperse; in past experiments, researchers showed that some types of music were effective at preventing excessive loitering.[24] Alternating between tailored music and focused, clear messaging might be a way to calm and entice a group to change its behavior, perhaps in part by engendering a degree of positive feelings.

---

[22] Michael Silver, *Tactics, Training, and Procedures for the Warfighter Reacting to Crowd Dynamics*, Anacapa Sciences, Inc., 2002.

[23] Mark Granovetter, "Threshold Models of Collective Behavior," *American Journal of Sociology*, Vol. 83, No. 6, May 1978; Mark Granovetter and Roland Soong, "Threshold Models of Diffusion and Collective Behavior," *Journal of Mathematical Sociology*, Vol. 9, No. 3, 1983.

[24] Lily E. Hirsch, *Music in American Crime Prevention and Punishment*, University of Michigan Press, 2012.

Peering into the Fog

Currently, the predominant theoretical framework that explains the behavior of large groups is social identity theory.[25] There are other characterizations, although research has debunked such "mob mentality" stereotypes that characterize individuals in a group as experiencing a sense of depersonalization and inordinate emotional reactions. An approach called *field theory* characterizes behavior as a function of both the person and their cultural environment. Another approach called *rational choice theory* assumes that people engage in thorough cost-benefit analyses to shape their choices; this theory depicts group members as rational individuals engaged in deliberative decisionmaking. Both theories tend to discount the role of emotions. However, in the stressful, complex, time-sensitive, incomplete-information situations that we are considering—especially the gray zone or combat—people might not always make fully informed, rational decisions.[26]

Social identity theory emphasizes the importance of acting similar to other group members to display one's own identity, which is also known as the *bandwagon effect*. Because social identity plays such a critical role in group behavior, those trying to anticipate and influence that behavior ideally would understand that group thoroughly, including what subgroups are present, as well as their motivations and goals, cultural norms, relationships, and leaders.

Unfortunately, when confronting a group of civilians, most operational and tactical commanders will almost always be operating with a high degree of uncertainty about many of these items. This limitation is not merely an issue of intelligence collection; group members and organizers themselves might have limited knowledge of how motivated individuals are or how they might respond to whatever they can perceive from their particular vantage points. Group members have limited visibility to see what is happening because of visual obstructions, including each other. How they perceive, comprehend, and project can depend heavily on their precise locations. As we noted previously, group members' decisionmaking will likely be affected by emotions, stress, and uncertainty, as well as by applying heuristics and biases rather than a rational set of calculations. Moreover, they will always have limited awareness of others' motivations and levels of commitment, and behaviors might spread or cease in ways that are hard for anyone to predict, including even organizers or agitators.

Research regarding the behavior of large groups of confrontational people is limited by lack of real-world and experimental data from which to draw conclusions. Various models have been developed, such as the Workbench for refining Rules of Engagement against Crowd Hostiles (WRENCH) simulation, which was developed by the Naval Postgraduate School and JIFCO.[27] WRENCH is a stochastic agent-based model that is designed to be a research tool and used to explore the effects of various security management approaches, particu-

---

[25] Drury and Reicher, 1999.

[26] Kurt Lewin, *Resolving Social Conflicts and Field Theory in Social Science*, American Psychological Association, 1997.

[27] Susan Aros, Anne Marie Baylouny, Deborah E. Gibbons, and Mary McDonald, "Toward Better Management of Potentially Hostile Crowds," paper presented at the 2021 Winter Simulation Conference, 2021.

larly regarding interactions with crowds of civilians, and WRENCH is useful when developing rules of engagement. However, WRENCH is not designed to inform decisionmaking in real time. It is unlikely that any agent-based models could be operationalized to be a real-time decision support tool because the necessary data about the demographics, motivations, movements, perceptions, comprehension, or projections of angry, road-blocking protesters most likely will not be available and incorporated into a model rapidly enough to help shape tactical decisions.

Similar problems arise when commanders are confronting opposing military forces, either in the gray zone or actual combat. Although intelligence can provide some information about the quality of their training and leadership, there might remain many known unknowns about how they will respond to NLW use and other behaviors. Our RAND colleagues have done research on military forces' will to fight, including some simulations of tactical encounters.[28] This research illustrates how much the will to fight can affect the outcome of military encounters, in addition to highlighting specifics, such as how a squad leader's rising levels of anxiety and anger could contribute to the squad fleeing from an encounter. The authors also highlight the dozens of variables that contribute to an individual's or unit's will to fight, including ideology, identity, and trust.[29] The details behind a specific unit's behavior, which is shaped by its commander and other personnel, will be unknowable in most cases. Even the will to fight of whole militaries can be difficult to judge; for example, the speed of the collapse of Afghan forces in 2021 and the ferocity of Ukrainian forces' resistance to Russia's invasion in 2022 were largely unexpected. Nonetheless, by enumerating what factors to consider, these studies lay the groundwork for future research, which might contribute to better future assessments of units' will to fight.

Closing Thoughts

In this chapter, we have briefly explored some factors that contribute to how individuals and groups behave, including how their behavior might be influenced by the use of NLWs. Judgments and decisions often employ mental shortcuts rather than rational analyses, particularly in stressful, information-limited, or time-constrained situations. Standard approaches to persuasion can often convince people to make particular choices that might not be rational. What people perceive, how they comprehend it, how they project the future, and how they decide to act are all influenced by limited information and environmental factors. Finally, even though commanders can make some assessments about how civilian groups or military units will behave using available data and might anticipate some ways to usefully influence

---

[28] Ben Connable, Michael J. McNerney, William Marcellino, Aaron B. Frank, Henry Hargrove, Marek N. Posard, S. Rebecca Zimmerman, Natasha Lander, Jasen J. Castillo, and James Sladden, *Will to Fight: Analyzing, Modeling, and Simulating the Will to Fight of Military Units*, RAND Corporation, RR-2341-A, 2018.

[29] Connable et al., 2018.

their behavior using NLWs or other capabilities, they will not know in advance precisely how NLWs will change the other party's behavior. In the next few chapters, these insights will be discussed in the context of the vignettes' different contexts.

CHAPTER 3

Overview of the Impact of NLWs and Vignettes

We begin this chapter by describing how activities employing NLWs contribute to DoD's goals, and then we provide a brief overview of the vignettes that will be explored in Chapters 4–6.

How NLWs Have Impact

We have previously developed a structure called a *logic model* that shows how activities conducted with NLWs lead to direct outputs, higher-level outcomes, and fulfillment of strategic goals derived from the National Defense Strategy.[1] Although this has been thoroughly explained in a prior RAND publication, we reproduce it here in Figure 3.1. Thick, dark lines indicate strong connections, while thin, light ones represent weaker ones.

Because it might be hard to trace some connections through the thicket of lines, readers may want to look at a dynamic version of this logic model that is available on RAND's website; this visualization tool enables a user to click on any element of the logic model and see which other elements are strongly linked to it.[2]

To walk through one example, when individuals with unclear intent are approaching military forces, those forces can use such NLWs as AHDs and laser dazzlers for the activity of revealing intent: Innocent civilians will generally back away, while attackers will likely keep approaching. This approach reduces the risk of accidentally harming civilians, which is represented by the output of minimizing collateral damage. In addition to the obvious humanitarian benefits, this contributes to the outcomes of avoiding the alienation of populations and

---

[1] James Mattis, *Summary of the 2018 National Defense Strategy: Sharpening the American Military's Competitive Edge*, U.S. Department of Defense, 2018. For more information on logic models, see Scott Savitz, Miriam Matthews, and Sarah Weilant, *Assessing Impact to Inform Decisions: A Toolkit on Measures for Policymakers*, RAND Corporation, TL-263-OSD, 2017.

[2] See Romita Grocholski, Savitz, Huerta, et al., 2023.

FIGURE 3.1
Logic Model for NLWs

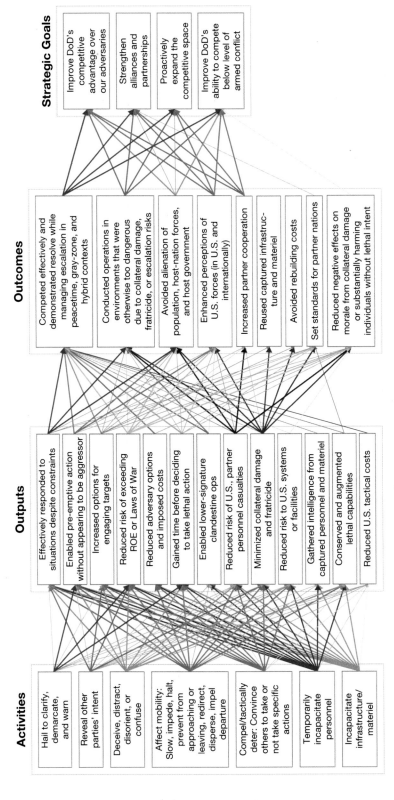

SOURCE: Reproduced from Romita Grocholski, Savitz, Litterer, et al., 2023, Figure 2.3.
NOTE: Thick, dark lines indicate strong connections, and thinner, lighter lines indicate weaker ones. All lines emanating from a single element are shown in the same color, to facilitate visual tracking of their common source; the colors have no other meaning, and similar colors in different columns are unrelated. ROE = rules of engagement.

governments, as well as enhancing perceptions of U.S. forces. Both of these outcomes contribute to the strategic goal of strengthening alliances and partnerships.[3]

Overview of Vignettes

We have developed a series of nine vignettes for analytical purposes and designed them to be varied: They address the sea, air, and land domains while also encompassing gray-zone, combat, and civilian encounters. They have also been developed partly based on real-world situations. These real-world examples include gray-zone confrontations in the sea and on land, such as between ships in the South China Sea or among infantry in the Himalayas. The confrontations in our vignettes also include urban combat, which emulates 21st-century fights in Ukraine and Iraq, and tunnel warfare, which is akin to parts of the ongoing war in Gaza at the time of this writing. One vignette highlights the recurring threat posed by uncrewed aerial vehicles (UAVs)—i.e., drones—such vehicles have become a mainstay of warfare in Ukraine, the Caucasus, the Red Sea, and elsewhere. Although the geography, names, and specific scenarios have been fictionalized for the reasons below, these vignettes reflect real-world challenges that military forces face.

The value in analyzing these vignettes is that they provide venues for exploring how NLWs and, in some cases, other IFCs could be used, including the advantages and disadvantages of particular tactics. Our vignettes also reveal gaps in current capabilities and highlight desirable future capabilities to develop or acquire.

For each vignette, we briefly describe the context, events, and assumptions that structure it, as well as the goals of the U.S. forces and any participating allies or partners. We then propose how NLWs could be used effectively and explain our assessment. We also describe potential risks and challenges, as well as desirable future capabilities.

Note that other than the United States, all country and geographic names are fictional. This was a deliberate choice so that readers would focus on the use of NLWs rather than the detailed specifics of the vignette or assessments of its plausibility. We do cite historical precedents that contribute to the credibility of each vignette. In addition, even though our vignettes presumably take place in a wider diplomatic context, our focus is exclusively on NLW employment within the vignettes rather than diplomatic coordination. For example, vignette D involves a NATO member, but we do not discuss potential invocation of NATO's Article 5—namely, that an attack on any member is an attack on all of them.

Table 3.1 provides a summary of the vignettes described in the next three chapters. To illustrate the diversity of the vignettes, we provide information on the domain (air, land, or sea) and context (gray zone, combat, or civilian) associated with each vignette. Each vignette also has a title and short description.

---

[3] More detail on these and other examples can be found in Romita Grocholski, Savitz, Litterer, et al., 2023.

TABLE 3.1

Summary of Vignettes with Fictional Countries

Domain	Context	Vignette	Description
Sea	Gray zone	A. Highway to the Dark Gray Zone	• A U.S. destroyer is conducting freedom-of-navigation operations in the West Hada Sea. • The destroyer is confronted by a Navy warship from the country of Akai that attempts to use eye-damaging lasers, electromagnetic warfare (EW), and ramming.
	Gray zone	B. The Ungoverned Country	• Warships from the United States and the adversary of Laal are both conducting non-combatant evacuation operations in confined, crowded waters around the country of Segol. • The Laal warships have attempted ramming and attacked their U.S. counterparts with eye-damaging lasers and EW. • Laal-flagged boats are repeatedly approaching U.S. warships despite warnings and then turning back.
	Gray zone	C. Rock, Paper, Symbol	• A destroyer from the country of Nyekundu has landed marines on an uninhabited island that the U.S. government recognizes as belonging to the U.S. ally of Midori. • U.S. and Midori destroyers want to get the Nyekundan forces to leave peacefully.
Land	Gray zone	D. Black Mountain and Gray Falcon	• A joint platoon from the nations of Rood and Gulabi has crossed over the border from Gulabi into NATO member Hara's territory. • U.S. and Hara forces have confronted the Rood and Gulabi forces and asked them to leave, but they are approaching and shouting insults.
	Combat	E. One Hundred Fires	• During a civil war in the country of Mera, U.S. soldiers and marines are helping rebels to take over the key coastal city of Stovatri. • U.S. and Meran rebel forces are trying to minimize civilian casualties.
	Combat	F. Descent into Hades	• A nonstate actor in the country of Lusam has launched an attack against Gudu, another country, killing and abducting many people, including U.S. tourists. • Many of the captives are being held in an extensive, well-secured tunnel complex in Lusam. • U.S. forces are supporting the fight and rescue effort.
	Civilian	G. Stymied and Imperiled	• U.S. soldiers are trying to move a convoy through the recently captured city of Bola in the country of Aparga, which was invaded by the dictatorship of Nuroni. • The convoy is blocked by a crowd of angry civilians and being targeted by gunfire from a building.
	Civilian	H. We Come to a Land Down Under	• After arriving at Lila, the country of Zold's main port city, U.S. marines are trying to move inland to the capital, Jamani, to help the government fight a rebel force. • Marine movements are impeded by a large number of displaced people fleeing the other way.

Table 3.1—Continued

Domain	Context	Vignette	Description
Air	Civilian	I. Drone, Drone of My Own	• Near an air base on the archipelagic nation of Inis, host-nation adolescents regularly launch recreational UAVs into the path of U.S. aircraft that are trying to land. • The Inis police are insufficiently responsive to arrest the adolescents, and U.S. officers lack jurisdiction.

Chapter 4 covers the vignettes that take place in the gray zone, Chapter 5 includes the vignettes involving combat, and Chapter 6 describes the vignettes involving large groups of civilians. We conclude each chapter with some high-level takeaways regarding its vignettes.

CHAPTER 4

Gray-Zone Vignettes

Gray-zone operations are those in which adversaries conduct operations to fulfill their goals without engaging in full-scale warfare. Such interactions occur regularly in various contexts. In this chapter, we explore four vignettes that illustrate how NLWs could be used in such gray-zone circumstances.

- **A. Highway to the Dark Gray Zone.** A U.S. destroyer is conducting freedom-of-navigation operations in the West Hada Sea and confronted by an Akai Navy warship that uses eye-damaging lasers and EW against the U.S. destroyer and has even attempted to ram it.
- **B. The Ungoverned Country.** Multiple countries, including the United States and its rival nation of Laal, are conducting non-combatant evacuation operations to get their nationals out of the rebellion-wracked country of Segol. In these confined, crowded waters, the Laal warships have attacked their U.S. counterparts with eye-damaging lasers and EW and have attempted to ram them. In addition, Laal-flagged boats are repeatedly approaching U.S. warships despite warnings.
- **C. Rock, Paper, Symbol.** A destroyer from the United States' rival nation of Nyekundu has landed its marines on an uninhabited island belonging to Midori, a U.S. ally. The rival marines have raised the Nyekundan flag and are contributing to an information campaign organized by Nyekundu to claim the island and others around it. Destroyers from the United States and Midori aim to get the Nyekundan marines to leave the island peacefully.
- **D. Black Mountain and Gray Falcon.** Forces from Gulabi and its ally, Rood, are conducting joint foot patrols along the Gulabi border with Hara, which is a member of NATO. The forces have crossed over the border and are ensconcing themselves in Hara's mountainous territory. A joint U.S.-Haran foot patrol has confronted them and asked them to leave, but personnel from Rood and Gulabi are closely approaching them while hurling insults.

Vignette A. Highway to the Dark Gray Zone

Context, Events, and Assumptions

In the disputed waters of the West Hada Sea, a map of which is shown in Figure 4.1, a U.S. destroyer is conducting freedom-of-navigation operations around a feature to assert international rights to do so. The West Hada Sea is a difficult area in which several nations contest territorial rights, and the country of Akai, a U.S. competitor, has claimed total control of vast swaths of the sea in contravention of international law. The confrontation in this vignette begins when a warship from Akai engages a U.S destroyer by aiming eye-damaging lasers (unlike the U.S. military's eye-safe laser dazzlers) at the destroyer's pilot house and aggressively maneuvering in ways that risk ramming it. The U.S. destroyer's personnel are wearing glasses to filter out the lasers, an effect which is bolstered by the ship's thick, tinted windows. The Akai warship refuses to engage in bridge-to-bridge communications, except to periodically demand that the U.S. destroyer depart.

This vignette is loosely based on events in the South China Sea. Ships from the People's Liberation Army Navy, the China Coast Guard, and the Maritime Militia have regularly con-

FIGURE 4.1
Map for Vignette A: Gray-Zone Encounter in the West Hada Sea

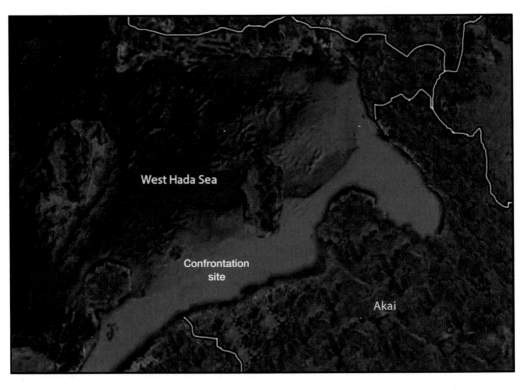

SOURCE: Firefly, output from a prompt by nhuer001 using Google Earth images, Adobe, January 5, 2024.
NOTE: Map and names are fictional.

fronted both military and civilian vessels from other nations, sometimes ramming them or threatening to do so. The frequency of these confrontations has intensified in the 2010s and 2020s, though such events repeatedly occurred in prior decades. U.S. ships have periodically conducted freedom-of-navigation operations around features in the South China Sea and sometimes sail through the Taiwan Strait; both are areas in which China claims sovereignty in contravention of international law. During such operations, U.S. ships have been threatened by their Chinese counterparts at times, which have sometimes come perilously close to ramming them.[1] Further afield in East Africa, Chinese forces have also used lasers to target U.S. military aircraft pilots in Djibouti.[2]

U.S. Goals

The U.S. military seeks to affirm freedom of navigation in the West Hada Sea and to demonstrate its commitment to maintaining this international norm. Moreover, the U.S. military seeks to deter future unlawful maritime claims by demonstrating the ability to counter them without needing to escalate or yield. This operation will also serve as a demonstration to the international community that the United States will firmly uphold international maritime law while avoiding unnecessary conflict.

Proposed Employment of NLWs

In response to the Akai warship's provocations, the U.S. destroyer is prepared to employ several IFCs. Initially, the U.S. destroyer could use LRADs or the experimental LIPE system that emits sound at a distance. Either or both could serve to communicate warnings and to generate irritating or distracting sounds to dissuade the Akai warship from continuing its provocations. As shown in Figure 3.1, key NLW activities include hailing, distracting, and achieving a measure of tactical deterrence. One factor contributing to deterrence is how the message is framed, which we discussed among the psychological insights in Chapter 2. The U.S. destroyer wants to structure its counterpart's choices in a way that is intended to favorably shape its decisions. The message is: If you keep behaving aggressively and trying to counter our peaceful and lawful behavior, we will continue to barrage you with unpleasant sounds that impair your ability to function and will escalate from there, so you should back off.

If the Akai warship does not change its behavior, eye-safe laser dazzlers could also be aimed at its bridge to impair and distract the personnel driving. EW tactics, such as jamming or spoofing communications and sensors, are another option. If these tactics are insufficient, the U.S. destroyer could potentially employ water cannons or send UAVs to buzz the Akai

---

[1] Ben Werner, "Destroyer USS Decatur Has Close Encounter with Chinese Warship," U.S. Naval Institute News, October 1, 2018; Luis Martinez, "Chinese Warship Cuts Off US Navy Ship, Marking 2nd Military Provocation in Week," ABC News, June 4, 2023.

[2] Aaron Mehta, "Two US Airmen Injured by Chinese Lasers in Djibouti, DoD Says," Defense News, May 3, 2018.

ship as a warning. Throughout the encounter, sensors would document the actions of the Akai vessel, to ensure that evidence is gathered to support U.S. actions and help shape the information environment.

Risks and Challenges

The main challenge is ensuring that the Akai Navy warship understands the non-lethal nature of the U.S. destroyer's tactics. If the Akai warship's captain misperceives U.S. actions as an act of lethal force—for example, if they think that the buzzing UAVs are explosive-laden—they might undertake lethal action in turn, leading to unwanted escalation and U.S. casualties. There might be some value in imbuing Akai forces with concern and uncertainty about what will happen next and not having them perceive that the U.S. military is more afraid of escalation than they are, but there is also a desire to avoid accidentally inflaming them with fear of imminent damage or loss of retaliatory capacity.[3] The inclinations of the Akai military to escalate and how this might be influenced by the use of NLWs are inherently unknowable in advance or during the encounter. Prior to this or similar operations, the U.S. military could communicate with Akai via back channels to try to minimize the risk of misperception. The U.S. military could also benefit from intelligence regarding Akai's overall perceptions of U.S. NLW employment through using that information to shape both communications and tactics.

There is also a challenge in managing the perceptions of the international community, even though the U.S. destroyer is thoroughly documenting the incident. Akai could disseminate false information about the encounter by using manipulated video or audio to suggest that the U.S. destroyer was initiating escalation through highly aggressive behavior. There is also a critical need to publicly communicate that the U.S. military is using eye-safe lasers that do not inflict lasting damage, whereas Akai is using lasers that permanently impair vision.

A key limitation is that some NLWs that would otherwise be desirable would not be effective in this context. Assuming that ADS were able to be employed from a ship (which has not been tried or tested), it would only work against personnel who are exposed on deck; the ship's extensive metal prevents the propagation of electromagnetic beams within it.[4] The ship can have most or all such personnel go below decks, and even if some remain exposed, it is unclear to what extent this might affect the captain's deliberations. The VIPER system that incapacitates the electronics of small vessels (which has not been employed on warships) would likely be ineffective against a large ship, and the MVSOT that impede propellers would likely have minimal effectiveness unless used in very large quantities.

---

[3] Sean Monaghan, *Deterring Hybrid Threats: Towards a Fifth Wave of Deterrence Theory and Practice*, European Centre of Excellence for Countering Hybrid Threats, Hybrid CoE Paper 12, March 2022; Thomas Schelling, *The Strategy of Conflict*, Harvard University Press, 1960.

[4] In electromagnetic terms, the ship's metal creates a Faraday cage, which is an enclosure that cancels out or blocks external electromagnetic fields, such as radio or microwave frequency–directed energy beams.

Desirable Future Capabilities

If technologically possible, it would be desirable to enhance capabilities to disable or impede large ships. One example might be a more capable MVSOT-like system that could deliver large quantities of material to impede even a massive ship's propeller. Unfortunately, given the amount of metal that surrounds a large ship's personnel and electronic systems, scaled-up versions of ADS and/or VIPER are unlikely to be able to have a substantial effect.

Vignette B. The Ungoverned Country

Context, Events, and Assumptions

In the country of Segol, a map of which is shown in Figure 4.2, a critical situation has emerged because of a massive rebellion and civil war. The Segol government has permitted warships from both the United States and a U.S. strategic competitor, Laal, to come and evacuate their citizens from the main commercial port. Laal warships have been aggressive against their U.S. Navy counterparts in the confined, crowded waters adjacent to the port. The warships

FIGURE 4.2
Map for Vignette B: Gray-Zone Encounter off the Coast of Segol

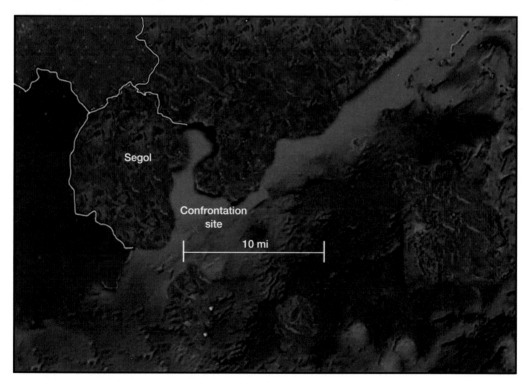

SOURCE: Firefly, output from a prompt by nhuer001 using Google Earth images, Adobe, January 5, 2024.
NOTE: Map and names are fictional.

have aimed eye-damaging lasers against pilot houses, and on multiple occasions, they have deliberately maneuvered in such a way as to risk ramming U.S. warships. (As in vignette A, U.S. sailors are wearing glasses to filter out the lasers, and some of the lasers' effect is attenuated by the thick, tinted windows of the ship.) The warship commanders have generally refused to engage in bridge-to-bridge communications during these interactions. In addition, small boats from the Laal Navy have also approached U.S. Navy warships, trying to provoke them by getting as close as they can without a lethal response.

Even though this vignette is mostly similar to vignette A, we want to explore whether confined waters and the contingency of a non-combatant evacuation would affect how NLWs might be used. We also introduce a situation involving small boats to see how similar or different this might be in terms of desirable NLW usage.

This vignette reflects the same real-world events in East Asia that were described in vignette A, as well as continual harassment of U.S. Navy and other vessels by Iranian small boats in the Persian Gulf. The U.S. military regularly conducts non-combatant evacuations when U.S. citizens or citizens of allied nations are in danger; for example, the U.S. Navy rescued people from Sudan during that nation's descent into civil war in April 2023.[5]

U.S. Goals

The primary objective of the United States is to enable its warships to protect themselves while they safely evacuate U.S. citizens.

Proposed Employment of NLWs

As in the previous vignette, U.S. Navy warships can counter threats from Laal warships using LRADs and/or LIPE systems for communication, deterrence, irritation, and impairment. As needed, U.S. warships can then employ eye-safe laser dazzlers. However, this confined, crowded environment would severely curtail the ability to employ EW, water cannons, and buzzing UAVs because of the risk of affecting third parties.

To counter the smaller boats, initial tactics would involve AHDs and/or LIPE, and then the use of laser dazzlers. Because the boat operators are not protected by vast amounts of metal, they can be further warned and impaired through the use of ADS to create a temporary heating sensation. Subjecting those personnel to a series of auditory, visual, and thermal stimuli could help the U.S. warship to project that continuing to approach would have ominous results. If this type of tactical deterrence fails, NLWs can perform another activity shown in Figure 3.1—namely, affecting mobility; the boats can be physically stopped through the use of VIPER to disable their electronics and/or MVSOT to impede their propellers.

---

[5] Mallory Shelbourne, "U.S. Navy Sends Nontraditional Ships to Support Sudan Evacuation," U.S. Naval Institute News, May 1, 2023.

Risks and Challenges

As in vignette A, there is a risk that the other nation's warships will choose to escalate despite U.S. attempts to manage the situation through the use of NLWs. It is desirable for Laal to have some fear and uncertainty regarding U.S. intentions, insofar as Laal might back off for fear of crossing red lines that will lead to escalation that it does not desire. However, precisely how Laal forces will react to NLW employment or what manner of NLW employment will convince them to back off rather than undertaking further aggression will always be imperfectly known.

There are also notably fewer NLW options against large ships that are desirable in this environment. The small boats are more easily addressed through various options that can keep them at bay without using blunt munitions or water cannons, both of which could risk harming personnel aboard the boats and leading to undesired escalation. Given the generally limited applicability of ADS in a maritime environment, it is not particularly likely that U.S. warships would cede the deck space needed for this system in its current configuration.

Desirable Future Capabilities

As was noted regarding vignette A, it would be desirable to have more non-lethal capabilities against large ships, including a more capable version of the MVSOT. Additionally, having a UAV-mounted version of ADS would be valuable.

Vignette C. Rock, Paper, Symbol

Context, Events, and Assumptions

A U.S. rival, Nyekundu, claims sovereignty over a tiny, uninhabited island that the U.S. government recognizes as belonging to Midori, a U.S. ally. The geographic area is shown in Figure 4.3. The island is important for both symbolic reasons and economic ones: Valuable fisheries surround it, and there might be substantial oil offshore. Although there is no human presence on the island, Midori's periodic patrols around it have helped to reinforce Midori's claims of sovereignty. Nyekundu has sent a warship to the area, which has launched small boats to land its marines on the island. Since their arrival, these marines have planted a Nyekundan flag on the island's highest point and have set up camp with enough basic equipment and supplies to sustain themselves for a few weeks. Meanwhile, the Nyekundans have released historic maps and documents on social media that provide "proof" that the island has always been part of their country. A U.S. amphibious ship and a Midori helicopter-carrying ship arrived in the area two days after the Nyekundan marines landed. In response, the Nyekundan warship has told the U.S. and Midori ships that they must leave the area and threaten to ram them.

In the interest of avoiding direct contact that could contribute to escalation, the United States and Midori have agreed that they will not land their own forces on the island; instead,

FIGURE 4.3
Map for Vignette C: Aiming to Protect Sovereignty over an Island

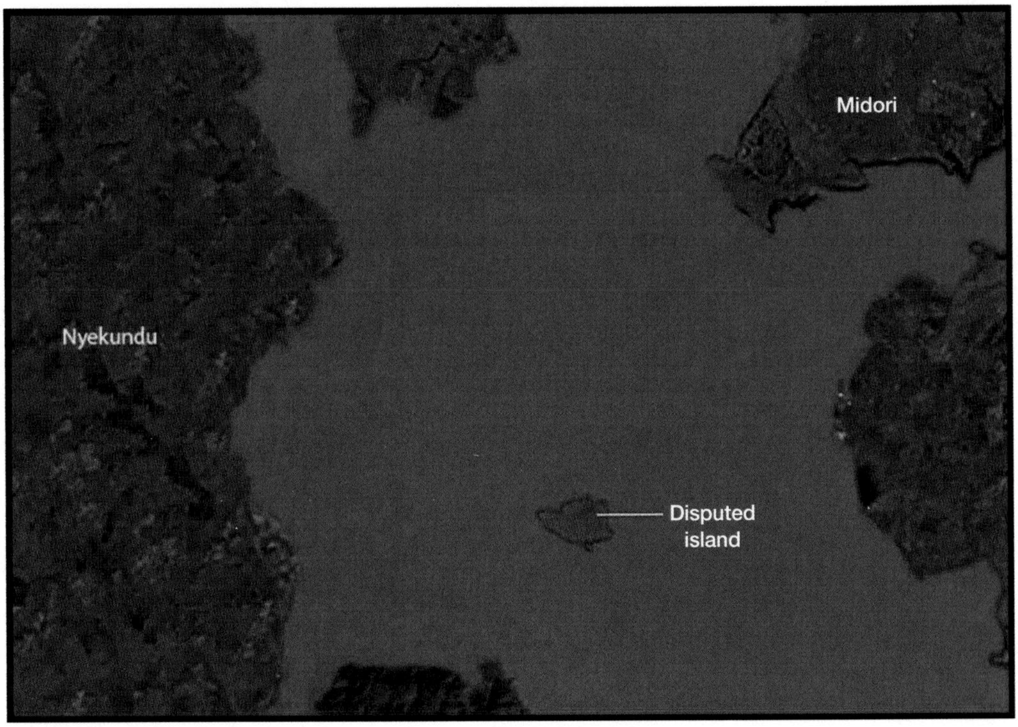

SOURCE: Firefly, output from a prompt by nhuer001 using Google Earth images, Adobe, January 5, 2024.
NOTE: Map and names are fictional.

they will pressure the Nyekundan marines to leave from offshore. They will also avoid using explosives or projectiles that could lead to rapid escalation.

This vignette is loosely derived from island sovereignty disputes in the East China Sea and South China Sea. As was noted in vignette A, although the frequency of maritime and island confrontations has intensified in the 2010s and 2020s, they have been occurring for a half century. Some of these have crossed the threshold to outright conflict: Chinese forces battled their Vietnamese counterparts to seize South China Sea islands during the 1970s and 1980s.

Even though the U.S. military has not worked with an ally to forcibly eject personnel occupying disputed islands, it is reasonable to envision that the occupation of islands that the U.S. government deems as belonging to an ally would be a red line for the United States, compelling it to either help or reverse the situation, lest its credibility as an ally and security guarantor be undermined.

U.S. and Allied Goals

The United States and Midori are concerned about Nyekundu establishing any type of presence on the island. They want to get the Nyekundan marines to leave the island and return to their ship peacefully without causing fatalities or injuries that could escalate the situation. They also want the Nyekundan warship to stop harassing their ships and depart the area. There is some time sensitivity, because it would be desirable to resolve the situation in favor of Midori and the United States before Nyekundu could potentially send additional ships.

Proposed Employment of NLWs

There are two aspects of the U.S. and Midori response to this situation in which NLWs might be employed: (1) countering the warship and (2) countering the marines. The same tactics described for vignette A would be used against the warship, with the limitation that their use would be restrained by the desire for the warship not to depart until the marines had reboarded it.

The marine-focused actions involve the U.S. amphibious ship using multiple sensory modalities to pressure the Nyekundan marines. Referencing the activities in Figure 3.1, the U.S. military is aiming to compel the Nyekundan marines to leave or, conversely, to tactically deter them from remaining. LRADs and/or LIPE can be used at ever-increasing volumes to tell the marines to leave the island and to irritate them with continual noises that wear down their morale and will to remain. Even though the marines can limit the effects of the noise with earplugs or headphones, these countermeasures also impose limitations on their own ability to communicate. In addition, loud noises can be felt from vibrations throughout the body, not only through the ears.

The noise could be supplemented by the intermittent aiming of laser dazzlers against the marines, repeatedly subjecting them to intense glare. This dazzling glare could contribute to sleep deprivation and overall exhaustion. Officers, non-commissioned officers (NCOs), and political officers (if Nyekundan forces have them) should be targeted most frequently with laser dazzler and ADS beams, because these individuals' exhaustion and frustration are most important in determining whether the marines leave or stay.[6] Overall, by inducing both continual and intermittent unpleasant sensations and by overstimulating the marines to the point of cognitive overload, the marines might be induced to make mental shortcuts that cause them to leave. Drawing on some of the psychological literature mentioned in Chapter 2, the U.S. amphibious ship can frame the choice in unequivocal, consistent terms: Depart the island and be left alone, or stay and continue to experience NLW effects. Although the targeted marines can be ordered to remain by their chain of command, recognition of their deteriorating state might cause higher-level officers to declare that they have already achieved victory by planting the flag and supposedly demonstrating Nyekundan sovereignty, enabling

---

[6] Some militaries, such as China's People's Liberation Army, have dual-command systems, in which political officers share authority over units with other officers.

the marines to leave while maintaining national pride. After they have departed, Midori could subsequently send a few personnel to the island to remove the offending flag and then depart immediately.

Further measures to frustrate and tire the marines could include the use of flash-bang grenades—potentially dropped by UAVs—to generate intense noise and light that temporarily impair some personnel. Swarms of small UAVs could also be employed during the day to create audio and visual stimuli, such as buzzing around above the marines' heads, repeating messages, and irritating sounds. Intermittently jamming or spoofing communications between the marines and their ship could impose further stress.

Risks and Challenges

U.S. and Midori forces are trying to influence the behavior of the Nyekundan marines without resorting to lethal force, in a context in which both sides know that it is unlikely to be used. Whether or when the marines depart will depend on whether their officers, NCOs, and political officers (and perhaps the chain of command) deem that they should, which depends on several factors, including the degree to which NLW effects exceed the marines' tolerance, their level of comfort in declaring that their presence to date constitutes success, any public real-time portrayal of the incident, and other diplomatic efforts. The extent to which NLWs will decisively shape this decision is unclear, and NLW usage should be coordinated with both diplomacy and information operations.

Desirable Future Capabilities

Simple modifications to current NLW capabilities would potentially be helpful in persuading the Nyekundan marines to give up their occupation of the island. For instance, mounting AHDs, laser dazzlers, or ADSs on UAVs would allow U.S. and allied forces to use these capabilities in closer proximity and more directly while maintaining their own standoff distances. Adding cameras to these UAVs could also prove to be useful in capturing footage for information operations campaigns.

Additional capabilities that would be relevant for the maritime environment and countering the Nyekundan warship are discussed above under vignette A.

Vignette D. Black Mountain and Gray Falcon

Context, Events, and Assumptions

The nation of Rood, which is shown on the map in Figure 4.4, is a strategic competitor of the United States and has been seeking opportunities to deploy its forces abroad. Rood has sent large numbers of troops to its ally, Gulabi, and the two nations are operating joint patrols along Gulabi's mountainous borders. One of those borders is with Hara, which is part of NATO. A joint Rood-Gulabi platoon has crossed the rugged border into Haran territory on

FIGURE 4.4

Map for Vignette D: Confrontation in the Black Mountains of Hara

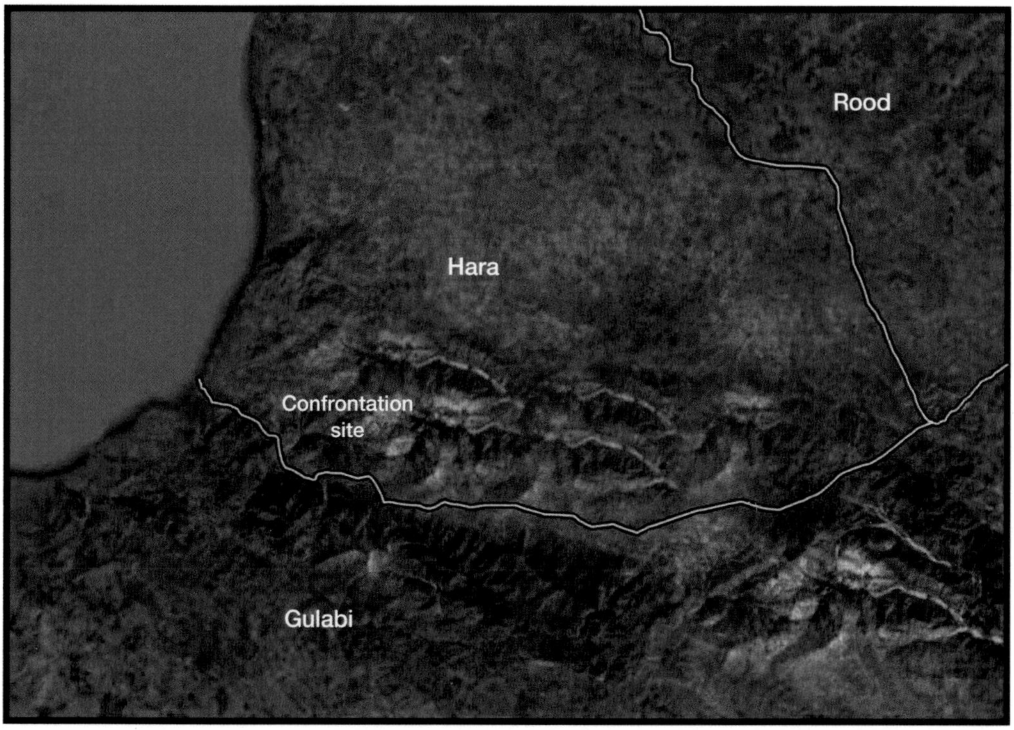

SOURCE: Firefly, output from a prompt by nhuer001 using Google Earth images, Adobe, January 5, 2024.
NOTE: Map and names are fictional.

foot and ensconced itself on a mountain within Hara, in a region known for its black cliffs and its magnificent falcons. The Rood and Gulabi soldiers have planted a large Gulabi flag showing its national symbol of a gray falcon, photographs of which they have uploaded to the internet via satellite communications. In response, a joint U.S.-Haran platoon has ascended the mountain on foot and told the Rood and Gulabi soldiers to leave. The U.S. and Haran forces have been met with refusal and various implausible, mutually inconsistent arguments. For example, the Rood and Gulabi soldiers claim that their GPS is faulty, that this place is universally recognized as Gulabi territory, that this land was stolen by Hara and they are reclaiming it for Gulabi, or that Hara is really not a nation at all but part of Gulabi. Most disconcertingly, they caution that the U.S. and Haran soldiers do not want to start a major war over a barren mountain.

Rood soldiers are trying to provoke the U.S. and Haran soldiers, insulting and closely approaching them, while Gulabi soldiers are laughing and capturing the action on video. The U.S. and Haran soldiers are being restrained thus far. Aside from the impact of this aggressive behavior on morale and information operations, it also poses a physical risk on a steep mountain if this degenerates into shoving or other physical altercations.

This situation is based partly on actual confrontations high in the Himalayas. Chinese and Indian forces who have met along their disputed border have not only shouted at and threatened the other side into leaving but have literally fought one another with sticks and stones, while also starting shoving matches and fistfights. Some of these confrontations along the craggy heights have resulted in fatalities.[7] Although such encounters have happened intermittently since the 1960s (and the two nations fought a full-scale war in 1962), the frequency and intensity of such face-to-face confrontations increased in the late 2010s and early 2020s.

U.S. and Allied Goals

The U.S. and Haran soldiers are trying to counter these provocations and force the Gulabi and Rood soldiers back over the border without degenerating the situation into outright fighting. Similar to the preceding vignette, this one involves the goal to compel or tactically deter, as shown in Figure 3.1. There is a clear interest in *not* incapacitating or immobilizing the other side for more than brief periods, because the overall goal is to get it to depart.

Proposed Employment of NLWs

The use of loud AHDs and laser dazzlers can help to overwhelm and distract the Rood and Gulabi soldiers and nudge them away from face-to-face confrontation. Although the wide-beam acoustic effects will cover a broad swath of hostile solders, the narrow-beam laser dazzlers can be aimed at specific individuals, notably unit leadership, such as officers, NCOs, and political officers, and the soldiers who are most aggressive.

In this close-contact environment, U.S. and Haran personnel will also be exposed to some of the irritating sounds, although they can potentially wear earplugs or other devices to make the sounds less intense (which the Rood and Gulabi forces also could do) at the cost of losing some auditory information. Depending on how the confrontation evolves, there might be times when it is desirable to have the AHDs emit music or other sounds that are likely to calm the other side rather than irritating or antagonizing it. Given the many known unknowns about how military units might behave, as described in Chapter 2, there will likely be an aspect of experimentation in using different NLWs and messaging to gauge the other side's willingness to step back from the brink.

As a very last resort, Tasers or blunt munitions could also be used to try to get the Rood and Gulabi personnel to back off. This option would likely require very high-level authorization.

---

[7] Gordon Fairclough, "India-China Border Standoff: High in the Mountains, Thousands of Troops Go Toe-to-Toe," *Wall Street Journal*, October 30, 2014; Jessie Yeung, "Indian and Chinese Troops Fight with Sticks and Bricks in Video," CNN, December 15, 2022; Arshad R. Zargar, "India-China Border Standoff Turns Deadly for First Time in Decades," CBS News, June 16, 2020.

Risks and Challenges

The physical proximity of the two sides makes it both urgent and difficult to employ NLWs in a way that induces the Rood and Gulabi forces to back off.

Desirable Future Capabilities

ADS could be desirable in this context by inducing uncomfortable heating sensations in various personnel, especially aggressive ones or unit leadership, to get them to back off. Unfortunately, ADS is not human-portable and requires substantial amounts of energy. For these reasons, ADS requires a vehicle, which is not available in this off-road environment. A lighter, battery-powered version of ADS, which could be carried by either personnel or UAVs, would be valuable in situations similar to the difficult terrain in this vignette.

Key Takeaways Regarding Gray-Zone Vignettes

There are several key patterns that are highlighted in the preceding vignettes. Some of these patterns are intuitive, but the vignettes illustrate them in ways that make these abstract points more tangible.

First, in maritime gray-zone standoffs involving large ships, there are limited differences between relatively open environments (such as in vignette A) and confined, crowded ones (such as in vignette B). In crowded waters, there might be some restrictions on how particular NLWs can be employed because of the risk of affecting other ships, but the general approaches remain the same.

Second, in the gray zone, there is a need to focus NLW capabilities on unit leadership. At the same time, NLW usage that affects others indirectly affects the commanders' decisions, and there might be value in targeting specific individuals who are particularly aggressive. However, because decisions are being made primarily by one or two individuals, these decisionmakers need to be targeted. This focused targeting is primarily an issue for NLWs that employ the electromagnetic spectrum, such as laser dazzlers and ADS. Although AHDs are directional, they can affect a wide swath of people at once, whereas the laser dazzlers and ADS have relatively narrow beams that require a focus on certain individuals.

Third, we encounter a key limitation of NLWs: They can impose physiological changes, but this might not always result in desired behaviors. Whether NLW effects are sufficient to change another party's overall behavior—beyond imposing some degree of incapacitation and discomfort—will depend on that party's degree of determination, which might also be influenced by unit culture, training, real-time public portrayal of a standoff, and ongoing diplomacy.

Finally, there are opportunities for additional NLW development that would be beneficial in situations illustrated by these vignettes and others similar to them. Items that would be desirable include a version of the MVSOT that could impede large ships, as well as AHDs or LARDs, laser dazzlers, and ADS that could be mounted on UAVs. Simply making ADS more portable and battery-powered would be a major step in the right direction. However, we have not analyzed the costs and technological hurdles that are associated with developing any of these technologies.

CHAPTER 5

Combat Vignettes

The two vignettes in this chapter take place in the context of combat.

- **E. One Hundred Fires.** During a civil war in the country of Mera, U.S. soldiers and marines are helping rebels to take over the key coastal city of Stovatri. The U.S. and Meran rebel forces are trying to minimize civilian casualties.
- **F. Descent into Hades.** A nonstate actor in the country of Lusam launched an attack in the neighboring country of Gudu, killing and abducting large numbers of people, including U.S. tourists. Many captives are being held in an extensive, well-secured tunnel complex in Lusam, and U.S. forces are supporting the fight and rescue effort.

Vignette E. One Hundred Fires

Context, Events, and Assumptions

The people in the nation of Mera, a map of which is shown in Figure 5.1, have rebelled against their dictatorship, which has been brutal to its own people and sponsors worldwide attacks against the United States and its allies. Defecting military units and local militias have waged a civil war and largely defeated the former government with U.S. assistance. One government holdout is the coastal city of Stovatri, whose population is divided between government and rebel supporters. The rebels and U.S. Army units are surrounding the city by land, and there has been a small landing of U.S. Marine Corps forces at an old port facility. As rebel and U.S. forces have advanced, they have been attacked by government forces during street encounters in which civilians have also often been present. In addition, groups of civilians in the streets have sometimes impeded movement.

The U.S.-rebel forces have set up a humanitarian corridor for civilians to escape the city and want those who are unable to leave to shelter in place and stay away from the streets. At the same time, landing craft resupplying the marine contingent are also being approached by small boats coming from the city, although it is unclear whether the people aboard the vessels are displaced people who are seeking help or hostile forces who are aiming to attack the landing craft.

This vignette draws on U.S. urban combat experiences alongside host-nation forces in Syria and Iraq during the 2000s and 2010s, including attempts to have civilians evacuate

FIGURE 5.1

FIGURE 5.1

Map for Vignette E: Meran Rebels and U.S. Forces Attempting to Capture the City of Stovatri

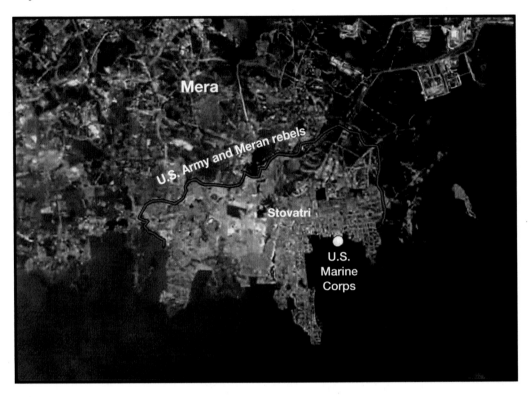

SOURCE: Firefly, output from a prompt by nhuer001 using Google Earth images, Adobe, January 5, 2024.
NOTE: Map and names are fictional.

before the most-intense fighting began. Notable examples include the Second Battle of Fallujah in 2004 and the capture of Raqqa, Syria, from the Islamic State in 2017.[1] There have also been instances of U.S. vessels being approached by others with ambiguous intent, such as an incident in which an Egyptian civilian was killed in the Suez Canal in 2009.[2]

U.S. and Allied Goals

The U.S. and rebel forces aim to capture the city of Stovatri while minimizing casualties among civilians and their own forces. U.S. and rebel forces want to keep civilians out of the paths along which they are advancing and to keep boats away from the landing craft. Minimizing

---

[1] Dexter Filkins and James Glanz, "With Airpower and Armor, Troops Enter Rebel-Held City," *New York Times*, November 8, 2004; "Syria War: Raqqa Deal Agreed to Evacuate Civilians," BBC, October 14, 2014.

[2] Voice of America, "US Confirms Egyptian Death in Suez Canal Shooting," November 1, 2009.

civilian casualties is not only a critical humanitarian concern but also important for ensuring the alignment of all rebel factions and popular support for the rebels. The U.S. and rebel forces can also employ NLWs in combination with lethal force against their adversaries, which will help to degrade their enemy's performance during combat. Furthermore, some NLWs can protect the flanks of movements through urban corridors.

In this context, key NLW activities (see Figure 3.1) will include hailing individuals to clarify and warn, revealing the intent of individuals who might be hostile or neutral, distracting adversary personnel, and affecting the mobility of possible flanking forces.

Proposed Employment of NLWs

There are several distinct components of NLW use in this context, such as protecting civilians, getting through the streets, and keeping the boats at bay. To protect civilians, it is critical to communicate about the humanitarian corridor and sheltering in place through various channels. These communication techniques include the usage of AHDs alongside traditional media (e.g., radio, television, pamphlets), social media, and internet announcements. Aerial signaling—e.g., using small UAVs to stream banners with messages or arrows indicating where the humanitarian corridor is—could also be valuable. AHDs can be deployed anywhere that U.S. and rebel platforms can operate on the ground, on the water, or in the air. Recorded rebel voices that use local dialects and accents are likely to be most convincing on AHD messaging and various other media.

To get through the streets, the U.S.-rebel forces can use AHDs to tell civilians to get away or stay away. AHDs, which have wide-area but directional effects, can also periodically emit irritating sounds to help accelerate departures. Uncooperative individuals who nonetheless appear to be civilians can be targeted intermittently with laser dazzlers and ADS. As we noted in Chapter 2, consistency of actions and messaging can help to persuade people to keep their distance.

At the same time, the advancing force can help to cover its flanks by using various non-lethal means of stopping vehicles. This strategy could be as simple as emplacing vehicles to block others or using mechanical solutions, such as spike strips and SNS–RDDs that puncture tires. A vehicle-mounted RFVS could also be employed to target vehicle electronics, whereas the PEVS, which is less expeditionary, would likely be less useful.

NLWs could also be used alongside lethal capabilities to scare, confuse, disorient, and impair enemy personnel. Unexpected stimuli alongside lethal fire could render soldiers ineffective or cause them to cut and run, despite being trained for combat. For example, the wide-area effects of irritating sounds could degrade their concentration and increase their anxiety, while targeting a few individuals with laser dazzlers or ADS could be disconcerting for the entire unit if members did not understand what was happening. Irritating noises, glare, and heat will make individuals less capable at shooting and distract them from being able to operate or communicate as effectively as they would otherwise. Even well-ensconced, trained snipers are not immune to the disruptive effects of loud sounds, glare afflicting their vision, and heating sensations of whatever parts of their bodies are not shielded by metal. In addition, using RFVS to halt adversary vehicles could also force personnel to dismount and disrupt their coordination.

As we noted in Chapter 2, people comprehend their perceptions based on prior experiences; in the absence of relevant experiences, they might assess the situation as worse than it is, which might cause them to fight faintheartedly or flee. Forces with prior awareness of U.S. NLW capabilities can be trained on how to employ countermeasures, such as special glasses or earplugs, and how operate while overcoming both NLW stimuli and the diminution of their senses from any countermeasures that they employed. However, combat is a terrible time to try to learn those skills.

Similar approaches could be used to keep boats away from the landing craft. AHDs could warn and irritate, thereby distinguishing intent, and those who continue despite warnings from AHDs are likely to be malevolent. Laser dazzlers could accentuate the message and confirm intent. If available, ADS could impair personnel, VIPER could disable the boats' electronics, and MVSOT could render the boats' propellers useless.

Risks and Challenges

Although all of the above can reduce civilian casualties, they are unlikely to eliminate them. The effectiveness of NLWs in impairing enemy soldiers or causing them to cut and run might also be diminished if those soldiers have been prepared to face them. For example, they might have already received training that inures them to the effects of NLWs, or they might have procured protective equipment that renders them less vulnerable, such as headphones that filter out some sounds.

Desirable Future Capabilities

A number of relatively simple technologies could be used to enhance the effects of NLWs. The ability to mount AHDs or other high-volume speakers on UAVs, as well as UAV-mounted signaling capabilities with pennants or lights, would facilitate the projection of messages throughout the city. Simple caltrops—metal objects shaped like toy jacks but with sharp points—could be tossed along forces' flanks to blow out vehicle tires. The ability to spray slick or sticky substances could impede personnel on foot and further protect flanks during urban movements.

Vignette F. Descent into Hades

Context, Events, and Assumptions

The country of Gudu, a map of which is shown in Figure 5.2, hosts a popular vacation spot for U.S. tourists and others, and this location is not far from Gudu's border with the country of Lusam. A nonstate actor based in Lusam launched a surprise attack against this vacation

FIGURE 5.2
Map for Vignette F: Hostage Situation in Lusam Tunnels

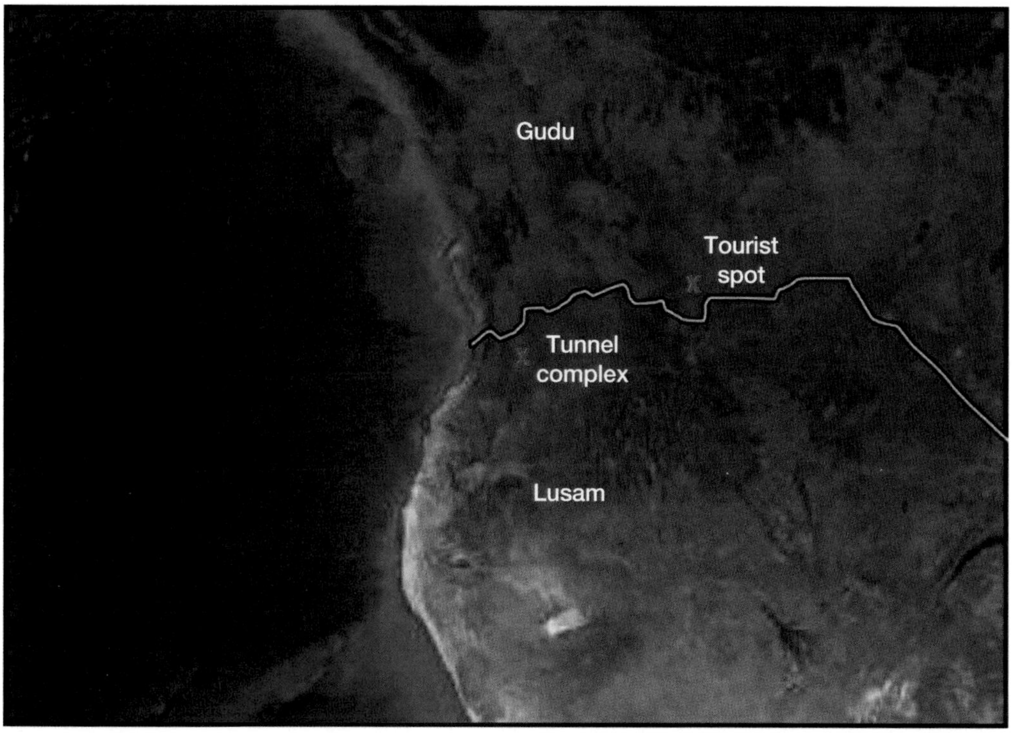

SOURCE: Firefly, output from a prompt by nhuer001 using Google Earth images, Adobe, January 5, 2024.
NOTE: Map and names are fictional.

locale, killing and abducting many people, including U.S. tourists. The captives were rapidly brought across the border to Lusam, where many of them are being held in a vast tunnel complex. U.S. forces, alongside counterparts from Gudu and other nations, are attempting to rescue the captives and strike the forces responsible for the attack.

This vignette, which involves the challenge of rescuing captives from tunnel complexes, reflects real-world events in the Israel-Hamas war in Gaza that began in October 2023.[3]

U.S. and Allied Goals

The goals are to rescue captives and to minimize risk to the U.S. and allied personnel entering the tunnels. This physical environment is particularly challenging because the tunnels' occupants have had the opportunity to design for defensibility. The tunnel occupants will also likely have emplaced various types of booby traps, created rooms from which they can

---

[3] David Ignatius, "Vast Gaza Tunnels Present a Battlefield of Nightmares," *Washington Post*, October 17, 2023.

shoot at intruders in the main passages while remaining almost invulnerable, and facilitated ambushes. The tunnels' occupants also have an informational advantage because they know the tunnels' precise layout. They might also have wired-in communications capabilities that facilitate coordination, but those entering the tunnels will largely lose access to the electromagnetic spectrum once they descend underground. Furthermore, the captives might be distributed in unknown locations throughout the complex, and a barely visible human might not be readily identifiable as a captive or a hostile actor, which makes the entire situation even more difficult.

Proposed Employment of NLWs

Directing annoying sounds from AHDs into tunnel entrances can keep people inside the tunnels awake, irritated, and exhausted. The focus is on temporarily incapacitating personnel (see Figure 3.1). Although this experience will be unpleasant for the captives, it might also build their confidence that the world has not forgotten them, and, most importantly, it can help to impair their captors. Even if the captors take to wearing headphones or other devices to counter the sounds, thereby limiting their auditory flow of information, sufficiently loud noises will reverberate through their bodies. A precedent for using such amplified sound exists: When deposed Panamanian dictator Manuel Noriega took refuge in the papal embassy during Operation Just Cause in 1990, intensely loud music was used to psychologically pressure him to depart that building.[4]

Changing the types of sounds over time to periodically cause captors to be alerted, only for nothing to happen, can make them complacent when a real incursion occurs. This strategy aligns with Cialdini's insight that people expect consistency, which we discussed in Chapter 2.[5] Aiming laser dazzlers into accessible, linear portions of tunnels and sending in UAVs or uncrewed ground vehicles (UGVs) can also cause adversaries to anticipate incursions that do not materialize. Those uncrewed vehicles can also collect valuable information about the tunnel complex and try to discover and detonate any booby traps that otherwise would have harmed humans.

When the time comes for an incursion, personnel who enter the tunnels would be armed with lethal capabilities but could also be given laser dazzlers, flash-bang grenades, Tasers, rubber-ball grenades, and other blunt munitions. The rescuing forces could employ these NLWs for situations in which they might have a couple of seconds or less to respond, during which time they may not be able to clearly distinguish the captives from their captors.

---

[4] Greg Myre, "How the U.S. Military Used Guns N' Roses to Make a Dictator Give Up," NPR, May 30, 2017.

[5] Cialdini, 1993.

Risks and Challenges

U.S. forces operating in such confined physical environments as tunnels might be constrained in terms of how much lethal and non-lethal gear they can carry. Moreover, the underground environment will be an extremely dangerous one in which forces will likely be information-starved. The adversary retains the ability to strike suddenly in unexpected ways and to use hostages as human shields. The adversary can even threaten to kill or harm the hostages in response to attempts to dislodge the captors with noise or other NLWs. Regardless of whether explicit threats are made, the sleep-deprived, frustrated captors might also be more inclined to torture, kill, or otherwise harm their hostages.

There are also legal strictures to consider. The continual use of loud noises for sleep deprivation, even against people who have the option of avoiding the noise by emerging to surrender, could potentially be considered torture or otherwise inhumane treatment. Legal review and assessment would be required before any such actions were undertaken.

Desirable Future Capabilities

Having uncrewed vehicles that could employ NLWs would be valuable; for example, a UAV or UGV entering a tunnel could employ ADS if that system were much smaller and battery-powered. In the near term, UAVs or UGVs could be mounted with already-compact NLWs, notably laser dazzlers or AHDs. They could also employ Tasers or similar electromuscular incapacitating systems, preferably those that have the ability to fire multiple "shots." Such UAVs or UGVs could immediately precede personnel entering the tunnels, contribute to the reconnaissance and detonation of booby traps as mentioned above, and impair adversaries in close proximity to the rescuers. Long, snaking wires linking UAVs and UGVs back to the tunnel entrance could be used for continuous feeds of information and remote control.

One capability that would potentially be desirable but is almost certainly precluded is the ability to use tear gas, pepper spray, or other incapacitating chemicals. Temporarily impairing or incapacitating everyone within the tunnels could facilitate rescue. For instance, Russia attempted to do something similar when it rescued hostages in a theater that had been seized by Chechen terrorists, but the incapacitating gas used apparently killed many of the hostages.[6] Rescuers in gas masks could then enter the tunnels more safely. Even if the hostage-takers have gas masks, they might not have as much training as the rescuers in operating while wearing them; gas masks impede vision and hearing in addition to requiring more effort to breathe. Moreover, although some gas masks have drinking tubes, there is no way to eat while wearing a gas mask. Over time, even masked hostage-takers might have to make a choice between exposure and weakness-inducing hunger. Another consideration is the risk to the hostages of prolonged exposure to some of these agents. The hostages would presumably have no protection from the chemical agents, and extended exposure to tear gas or pepper spray

---

[6] Artem Krechetnikov, "Moscow Theatre Siege: Questions Remain Unanswered," BBC, October 24, 2012.

could cause respiratory problems or even death, particularly for children, older people, the infirm, or people who had been weakened by their ordeal.

Regardless of the tactical desirability of using incapacitating chemicals, it might be precluded by international and U.S. laws. The Chemical Weapons Convention—which the United States has ratified—forbids the use of temporarily incapacitating agents in war.[7] However, such chemicals are permissible in the context of domestic policing. The use of such chemical agents by the military is also forbidden under Executive Order 11850, which was issued by President General Ford in 1975, with several exemptions. The President can approve the use of chemical agents when civilians are being used as human shields and such agents would save civilian lives, when the military is conducting rescue missions in isolated areas, or when military movements in non-combat areas require protection from terrorist attacks.[8] It is possible that these caveats, combined with questions regarding whether this action constitutes a "war" or a rescue, might permit the use of such capabilities.

Key Takeaways Regarding Combat Vignettes

Our biggest finding from exploration of these vignettes is that non-lethal capabilities can be useful in combat in addition to the gray zone, which we covered in the previous chapter, and civilian encounters, which we will cover in the next chapter. Notably, NLWs can be employed effectively in complex urban and tunnel environments that include non-combatant populations. NLWs can be used to help remove civilians from the combat environment and reduce the risk of those who remain. The introduction of irritating or disconcerting stimuli can both impair adversary forces and potentially lead them to flee or surrender. In hostage situations, fatiguing and incapacitating captors is highly desirable.

As the Ukraine-Russia and Israel-Gaza conflicts have illustrated, combat in the 21st century sometimes occurs in complex urban and tunnel environments that include civilian populations. NLWs can help U.S. forces to achieve their aims in these environments and limit casualties of both soldiers and civilians.

---

[7] Convention on the Prohibition of the Development, Production, Stockpiling and Use of Chemical Weapons and on Their Destruction, 1992.

[8] Executive Order 11850, 1975.

CHAPTER 6

Civilian Vignettes

The three vignettes in this chapter take place in the context of encounters with groups of civilians.

- **G. Stymied and Imperiled.** U.S. soldiers are trying to move a convoy through the recently captured city of Bola. The convoy is blocked by a crowd of angry civilians and being targeted by gunfire from a nearby building.
- **H. We Come to a Land Down Under.** After arriving at Lila, the country of Zold's main port city, U.S. marines are trying to move inland to the capital, Jamani, to help the government fight a rebel force. U.S. forces are impeded by many displaced people who are fleeing the conflict.
- **I. Drone, Drone of My Own.** Adolescents are regularly launching large numbers of recreational UAVs into the path of U.S. aircraft that are trying to land at an air base on the archipelagic nation of Inis. The local police are insufficiently responsive to arrest them, and U.S. officers lack jurisdiction.

Vignette G. Stymied and Imperiled

Context, Events, and Assumptions

The dictatorship of Nuroni has launched an invasion of Aparga, a U.S. ally, the locations of which are shown in Figure 6.1. U.S. and Apargan forces have defeated the original incursion and pushed into Nuroni's own territory. U.S. soldiers and marines are attempting to establish control over the port city of Bola on Nuroni's east coast. Bola is home to about 350,000 people who are known to be highly anti-American. No large-scale Nuroni units remain within Bola, but a few Nuroni soldiers are still there and periodically shoot at U.S. forces or throw grenades from rooftops and windows. There is evidence that some civilians—or at least people who are not wearing military uniforms—have also participated in these aggressive activities.

A U.S. convoy that is working its way through a densely packed portion of downtown Bola finds itself facing a massive group of a thousand or more civilians blocking its path. The convoy is stuck and has almost no ability to maneuver through the narrow streets as it tries to avoid running over civilians. Moments later, snipers begin shooting at the convoy from the

FIGURE 6.1

Map for Vignette G: Moving Through the Recently Captured City of Bola

SOURCE: Firefly, output from a prompt by nhuer001 using Google Earth images, Adobe, January 5, 2024.
NOTE: Map and names are fictional.

higher floors of a multistory building along the street. Civilians are also leaning out of windows in that building and attempting to see what is happening.

This vignette is based on real-world experiences of convoys that have been trapped in environments in which vehicles are confined to narrow roads. This scenario occurred in the U.S. intervention in Somalia in 1993, when militia barricades blocked convoys, and more recently in Ukraine, where lead vehicles were targeted beginning in 2022.[1] Snipers ensconced in tall buildings are also a mainstay of urban combat.

U.S. and Allied Goals

The U.S. and Apargan forces have two goals: (1) Neutralize the snipers, and (2) get the group to disperse. They want to accomplish these goals while minimizing civilian casualties. Accord-

---

[1] Mark Bowden, "The Legacy of Black Hawk Down," *Smithsonian Magazine*, January 2019; David Axe, "Hit the First Vehicle, Hit the Last & Trap the Rest: The Ukrainians Used a Classic Tactic to Devastate a Russian Ammo Convoy," *Forbes*, November 14, 2023.

ing to the logic model in Figure 3.1, NLWs can help to hail and warn, reveal intent, distract, affect mobility, and compel departure.

Proposed Employment of NLWs

The snipers will be targeted with lethal force (i.e., the U.S. and Apargan forces will shoot back). In parallel, the convoy could use AHDs to project loud, irritating sounds at the building's windows. This amplified noise will both impair the snipers and help to convince curious civilians to back away from windows. Handheld laser dazzlers could be aimed at individual windows to complement the sounds, which would further impair the snipers and discourage curious spectators. The combination of intense sounds and glare would presumably degrade even the most determined sniper's ability to accurately target the convoy. ADSs could conceivably be used for the same purpose, but because such systems are bulky, expensive, and likely to be few in number, they could only affect a small number of windows at any given time. An unfortunate property of both laser dazzlers and ADS is that their relatively narrow beams would preclude any one device from covering a wide portion of the building. Regardless, some combination of sounds, lights, and ADS would likely impair snipers and reduce the number of bystanders in the windows. Admittedly, some civilians might try to persevere to watch what is happening, while others might be forced by the snipers to serve as human shields, but any reduction in the number of civilian spectators is advantageous.

In parallel, AHDs and/or LIPE could also project sound to communicate with and disperse the group on the ground. The ability to project sounds primarily toward the back of the group, using LIPE or UAV-based speakers, would be valuable; the people farthest from the convoy are the ones that need to disperse first to avoid a possible stampede that could kill people and would trap the convoy. Messages could also be shared using visual signals, such as banners towed by UAVs. Regardless of how these messages in the local language are shared, they would tell people about other reasons to depart in addition to their own safety, such as an abundance of supplies at a nearby location. Projecting calming, reassuring sounds might help to make the group on the ground less belligerent, even as those in the windows above them are being subjected to sounds that irritate and impair.

If a few individuals or confined groups are particularly belligerent despite these efforts, they can be targeted with narrow-beam NLWs—namely, laser dazzlers and ADS—to briefly impair these people. It is important to *not* impair the vast majority of people on the ground, whose ability to voluntarily migrate is critical to the convoy's ability to advance. The departure of some people might foster departure of most of the rest, a psychological consideration that we discussed in Chapter 2. If the measures above fail to cause most of the group to depart, continual ADS usage could potentially open up a corridor by getting enough people to move to either side of the beam.

After most of the group recedes, the much smaller number of uncooperative individuals can be targeted with irritating sounds to drive them away. Laser dazzlers and ADS can be aimed at any individuals behaving aggressively toward the convoy or at those who seem to be encouraging others to remain in place. Those who still refuse to disperse can then be targeted using paintball rounds to mark them, flash-bang grenades to impair them, or blunt munitions, such as rubber bullets, that might temporarily incapacitate them. Depending on the circumstances, lethal force might be authorized in some cases.

Risks and Challenges

This scenario is a case in which very different types of NLW usage need to be employed in parallel against two distinct groups. Inevitably, there may be interference; the people in the crowd below will hear some of the sound projected at the upper floors of the building. There also might be competing demands for NLW capacity at the same time, forcing commanders to select where to direct NLW efforts.

Desirable Future Capabilities

The experimental LIPE would be useful in projecting sound at a distance, so that the most distant part of the group is the one that is most exposed to the message at the outset. Similar effects could also be achieved through the use of UAVs with loudspeakers. It would also be desirable to have automated ways of aiming laser dazzlers at targets, so that a device could maintain control of an array of dazzlers aimed at numerous windows. UAV-mounted laser dazzlers, AHDs, and even ADS (if it can be made more compact and less energy-intensive) could be used to project effects precisely where they are desired from an elevated vantage point. Wider ADS beams would also be desirable, if technologically feasible, to persistently cover multiple windows or groups within the crowd.

Vignette H. We Come to a Land Down Under

Context, Events, and Assumptions

There has been a coup attempt in the historically peaceful country of Zold, a map of which is shown in Figure 6.2. Zold sits a bit south of the equator and is known for its expansive, windy deserts. The coup has caused a civil war in which the United States and other nations are intervening to support the recognized government. A U.S. Navy amphibious ship has docked in Lila, Zold's second-largest city and main port, which is still under government control. The U.S. marines and their vehicles have disembarked and are heading toward the capital, Jamani, where most of the fighting is occurring.

Meanwhile, tens of thousands of displaced civilians are walking from Jamani toward the coast across roughly 150 miles of desert. Some have been walking for over a week and are taking frequent breaks or camping by the side of the road, which is blocking passage. These people are hungry, thirsty, and tired; calling out to those around them; and asking for food, water, shade, and medical care. A few of the small towns along the way have experienced theft

FIGURE 6.2

Map for Vignette H: Marines Head for Lila While Displaced People Flow the Other Way

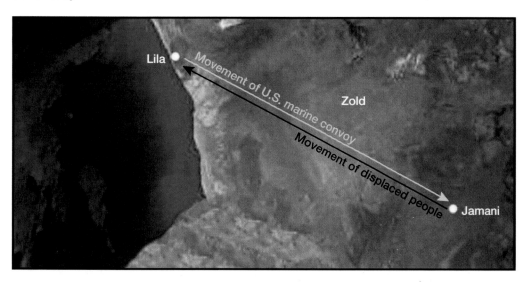

SOURCE: Firefly, output from a prompt by nhuer001 using Google Earth images, Adobe, January 5, 2024.
NOTE: Map and names are fictional.

and looting, and there have been a few fights among the displaced civilians and those community members.

The U.S. marine convoy has effectively been stopped by these waves of people walking the other way and by groups of civilians who have merely stopped. Both sides are frustrated with the other for blocking their respective path. This aggravation has resulted in a few incidents of rock-throwing and mobs surrounding the vehicles in the convoy. The marines have received intelligence that some supporters of the coup are mixed in with the displaced civilians. These detractors are intent on stirring up trouble to delay the U.S. forces' arrival in Jamani and inciting incidents between the marines and civilians to create hostility toward the U.S. military and the recognized government.

Although this vignette has some similarities with vignette G—both vignettes entail trying to move forces through a corridor despite impedance by civilians—there are also key differences. First, the physical environment is very different. Traversing a road through the desert, where people can easily move to one side, is very different from moving through a crowded urban environment. The people in question are not massed but distributed. Second, there is no active shooting, which simplifies the problem and reduces time-criticality.

This vignette reflects the fact that U.S. forces have dealt with columns of refugees fleeing in the opposite direction of their movements before, notably during the Korean War (1950–1953).

U.S. and Allied Goals

The goal of the U.S. marine convoy is to get to Jamani as quickly as possible to lend support to the sitting government of Zold, which requires getting civilians off the road. The U.S. forces want to do this while minimizing any risks to civilians and maximizing positive interactions with them to retain the populace's support for the current government and U.S. involvement and aid. The key NLW activities from the logic model (see Figure 3.1) are hailing to clarify, demarcate, and warn and revealing intent as needed. In some cases, there might also be a need to affect individuals' mobility if they pose a threat.

Proposed Employment of NLWs

In this vignette, the purpose of NLWs is primarily to communicate with and entice the displaced people to leave the road and move to a safe distance to one side. AHDs, physical signs, banners towed by UAVs, and other methods can be used to communicate the need to move for safety reasons. Multiple NLWs with different sensory modalities might help to overcome the bright sunlight, dust storms, and strong winds that might intermittently impede the usage of visual and auditory NLWs. If possible, this effort can be supplemented through the creation of occasional aid stations providing shade, water, and food at points along the road. Projecting the sound of local music might help to soothe angry tempers and encourage people to go to the aid stations. As we noted in Chapter 2, shaping perceptions in ways that local populations can easily comprehend and that foster mutual liking can make the marines' message more persuasive.

Any individuals who refuse to depart the road despite such enticements and messaging can be targeted with irritating sounds, laser dazzlers, and/or vehicle-mounted ADSs to the extent that environmental conditions permit. If any vehicles are approaching along the road, one non-lethal mechanism for stopping them is an RFVS to disable their electronics. Drivers can also be warned with AHDs and targeted with laser dazzlers to induce them to stop. As we noted in the previous vignette, continual use of ADS could even create a corridor, as a line of people would actively seek to move to either side of the beam.

If there are any individuals who threaten the convoy or continue to impede its movement despite these measures, then flash-bang grenades, blunt munitions, Tasers, or even lethal force might be appropriate NLWs to employ.

Risks and Challenges

Desperate people who are exhausted, thirsty, and hungry might be aggressive without thinking through the consequences. This is why repeated, mutually reinforcing messaging and distribution of aid are critical to garnering their cooperation. The marines want to avoid conflict that could result in civilian casualties, make the population more hostile, and delay movement toward the capital.

Desirable Future Capabilities

This vignette can be mostly carried out with existing capabilities, although there might be some additional utility to having AHDs mounted on UAVs, so that the convoy could communicate with people or irritate them with sounds at greater ranges. It would also be advantageous to have a vehicle-mounted ADS with a relatively wide beam to create a corridor, as needed.

Vignette I. Drone, Drone of My Own

Context, Events, and Assumptions

At a remote U.S. air base on the archipelagic country of Inis, the map of which is shown in Figure 6.3, a few dozen local adolescents regularly launch large numbers of recreational UAVs into the path of U.S. aircraft that are trying to land and take off. Despite the permissive use of EW and nets to take down the UAVs, the problem persists. Furthermore, there are also

FIGURE 6.3

Map for Vignette I: Adolescents Putting UAVs in Path of U.S. Aircraft Landing on Archipelagic Nation of Inis

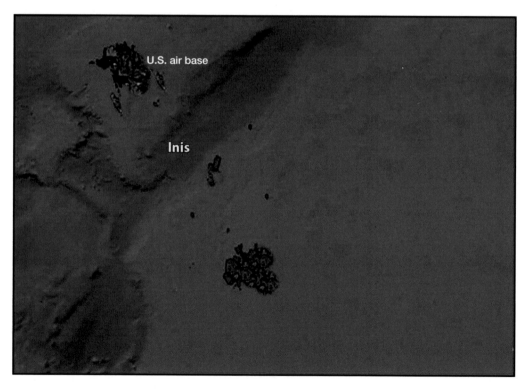

SOURCE: Firefly, output from a prompt by nhuer001 using Google Earth images, Adobe, January 5, 2024.
NOTE: Map and names are fictional.

indications that a hostile government funds these activities by providing commercial UAVs for this purpose and giving pocket money to the adolescents. The U.S. Air Force has routinely been able to localize the groups of adolescents controlling the UAVs by using its own overhead capabilities. However, when it has informed the local police in Inis, the adolescents have scattered and disappeared before the police can apprehend them. Local Inis authorities have approved the use of non-lethal force by U.S. forces beyond the base boundary but not lethal force.

This vignette draws on the fact that UAVs have been used to interfere with aircraft takeoff and landing. For example, UAVs operating around London's Gatwick Airport compelled its closure for two days in 2018.[2]

U.S. and Allied Goals

The goals of the U.S. air base personnel and local Inis police are to impair the adolescent controllers and impede their movements or simply to discourage and drive them away when they appear to be setting up their UAV launches. In terms of the activities listed in Figure 3.1, NLWs will be used to distract and affect mobility.

Risks and Challenges

Because the UAV controllers are minors and nationals of Inis, the U.S. personnel need to be especially careful not to inflict any type of enduring harm. The mere perception that these individuals are being maltreated by U.S. forces could result in loss of basing privileges.

Proposed Employment of NLWs and Desirable Future Capabilities

Current NLW capabilities do not work well in this context, but some UAV-based adaptations of them might. UAVs that could employ laser dazzlers—some of which are small enough to fit inside a palm—would be useful for rapidly impairing controllers. Suddenly experiencing a blinding glare might also cause them to stop trying to control the vehicle and perhaps even diminish their desire to conduct future attacks. Another system that would be desirable is a UAV that could spray a slick or sticky substance on the area around the controllers, hindering their movements to facilitate apprehension. UAVs that could launch paintballs to mark the perpetrators or that use compact versions of ADS to briefly impair them and discourage them from repeating this behavior could also be effective.

[2] Simon Calder, "Five Years on from Gatwick Drone Disruption, What Happened—and Could It Happen Again?" *Independent*, December 19, 2023.

Key Takeaways Regarding Civilian Vignettes

There are four key takeaways that emerged from the civilian vignettes. First, as we saw in the differences between vignette H and vignette I, the physical context matters immensely when trying to use NLWs to affect the movements of civilian groups. Civilians blocking a road in a confined urban environment will require different approaches for NLW employment compared with dealing with civilians blocking a road through a desert. Environmental conditions also influence the effects of auditory or visual stimuli—i.e., it might be harder to see or hear amid dust storms and strong winds.

Second, even though the precise motivations of groups of civilians might be unclear, a limited understanding of overall group disposition can help to shape response. The extent to which NLWs should be used to dissuade, impair, or encourage particular behaviors depends partly on an awareness of what the groups are trying to achieve through their movements and actions.

Third, in some environments, it could be advantageous to have additional ways to non-lethally and persistently cover lots of aimpoints. For example, it would be valuable if laser dazzlers or ADS could persistently contribute to suppression of snipers in windows, alongside AHDs.

Finally, we noted a number of desirable NLW capabilities. These include UAVs that carry speakers, laser dazzlers, or ADS; UAVs that create visual effects (e.g., by towing banners); UAVs that can disperse slippery or sticky substances; automated ways of aiming laser dazzlers at targets; and wider ADS beams.

CHAPTER 7

Key Findings and Recommendations

Key Findings

NLWs Are Effective in a Range of Operational Contexts

Our analysis of nine vignettes illustrates that NLW employment can assist with meeting operational and strategic goals across a range of contexts, particularly in complex environments (e.g., urban centers) and other environments where civilians are present. These situations include combat and other scenarios in which lethal force is also being used: NLWs can not only help to keep civilians from getting harmed but can even degrade the military performance of hostile forces by subjecting them to unexpected and distracting stimuli. It is even possible that such stimuli could contribute to flight or surrender. Commanders and planners should think of NLWs as another support arm, one that can contribute a unique set of effects to help accomplish operational objectives, particularly in situations in which the employment of lethal weapons is constrained or could have negative consequences, such as unintended escalation or lost basing privileges.

The vignettes also showed that two NLWs were useful in virtually all of the diverse contexts that we explored. AHDs and laser dazzlers were employed in eight out of nine vignettes, serving with versatility as communications tools, distractions, and means of inflicting temporary impairment. Whether they are separate systems or combined in HALLTS or EoF CROWS, sending direct acoustic and visual signals seems to be one of the most useful approaches.

Despite the broad utility of NLWs, there were a few contexts in which current NLW capabilities had limited impacts or were not useful. In vignette I, none of the existing NLW capabilities was applicable to addressing the civilians using small commercial drones to disrupt activities at an air base. Maritime environments involving larger adversary ships, such as vignettes A, B, and C, also illustrated that there are few current NLW capabilities against large ships.

Key Factors Influencing the Effectiveness of NLWs

To maximize the effectiveness of NLWs, some broad sets of factors need to be taken into account:

- **Psychological factors and group dynamics.** Effective use of NLWs requires that the results of human-effect research regarding NLWs be combined with research regarding psychology and group dynamics. People are generally responding to highly incomplete and often uncertain sensory perceptions, which people comprehend through the lenses of their experiences and emotions, use to project the future, and reference to make decisions. Decisions are often made using mental shortcuts called *heuristics*, and the decisions that are made can be powerfully influenced by the way the situation is framed or by initial data points. Decisions can also be influenced through well-honed approaches to persuasion, such as demonstrating consistency. Although there is always some unpredictability about the behaviors of military units and large groups of civilians, NLWs can be used to influence the behaviors of individuals within those groups—particularly leaders—in ways that cause other group members to alter their behavior accordingly.

- **The impact of the operational environment.** NLWs can be particularly useful in confined, crowded, or complex environments, such as urban centers and tunnels. However, as we saw in vignette E, very close ranges can cause some NLWs (e.g., AHDs) to affect U.S. forces in addition to the desired populations. Similarly, we saw in the contrast between vignettes A and B that the use of some capabilities that would be desirable in an open-water maritime context would be more constrained in a crowded, confined one. Because the environment determines the types of transportation available, it also determines what types of NLWs can be used—e.g., the bulky, power-intensive ADS is only viable for infantry members who are accompanied by a vehicle. Finally, the physical environment can directly affect the extent to which NLWs achieve their desired sensory effects. As we saw in vignette H, for instance, dust storms and loud, intense winds can impede auditory and visual signals.

- **The potential for countermeasures.** As we saw in vignettes E and F, NLWs are not immune to the measure-countermeasure dynamic that is pervasive in human conflict. People who expect specific NLWs to be used might be able to counter them with equipment as simple as earplugs or special glasses. However, this does not negate the value of NLWs. Countermeasures only partly offset the effects of NLWs, and some NLWs lack ready countermeasures that can be used against them. Employing countermeasures diminishes individuals' capability to perceive sensory information, degrading their capabilities. Even if an opposing military force has countermeasures available, it might not have trained collectively to operate effectively while using them. Even though countermeasures need to be taken into account, particularly in repeated encounters between U.S. forces and specific groups, NLWs can still have an impact.

Desirable Future Capabilities

Our analysis revealed that investment in some potential future capabilities might help increase the impact and versatility of NLWs and can have an impact in varied operational scenarios. The desirability of employing NLWs aboard UAVs emerged across five of the nine vignettes and sometimes in multiple contexts within a vignette. UAVs were used as communications tools and to apply the effects of NLWs more precisely. UAVs were particularly helpful in dealing with confined physical environments and large groups of people and also enabled the use of NLWs at longer ranges than is currently possible.

Additional future capabilities that were desirable in our vignettes included MVSOT-like capabilities that could be employed against large ships, automated systems that could point multiple laser dazzlers at multiple moving or stationary targets, and the more portable, battery-powered ADS, potentially with a wider beam.

Recommendations

Our overarching recommendations are for JICFO to

- **incorporate psychological, group-dynamic, and human-effect insights** into plans, concepts of employment (CONEMPS), and tactics for using NLWs that affect human beings and their behavior, including consideration of adversary perspectives and approaches to communicating that NLWs' effects are reversible[1]
- **prioritize the highly versatile AHDs and laser dazzlers** for service acquisition, training, and fielding, ensuring that both are highly integrated into operations
- **pursue such desirable future capabilities as UAV-based NLWs, MVSOT-like capabilities against large ships, automated aiming of laser dazzlers, and a smaller, lighter ADS**
- **ensure legal review** of NLW CONEMPS and tactics because, in some cases, employment of NLWs might be impeded by legal strictures
- **get more data** to support service-specific development of tactics and CONEMPS for NLW usage via modeling, wargaming, live exercises, and use in real-world operations, as appropriate.[2]

---

[1] The effects of antimateriel NLWs, as well as physiologically incapacitating NLWs such as Tasers, are more readily understood without drawing on psychological and group-dynamic research. However, it can be useful to analyze the psychological or group-dynamic effects of incapacitating materiel or other individuals.

[2] Modeling that incorporates psychological and group-dynamic principles could be particularly valuable, such as the WRENCH model developed by the Naval Postgraduate School and JIFCO. See Aros et al., 2021; and Susan Aros and Mary McDonald, "Simulating Civil Security Activities in Stability Operations," paper presented at the Interservice/Industry Training, Simulation, and Education Conference, November 2023.

Abbreviations

ADS	Active Denial System
AHD	acoustic hailing device
CONEMPS	concepts of employment
DoD	U.S. Department of Defense
EoF CROWS	Escalation of Force Common Remotely Operated Weapons Station
EW	electromagnetic warfare
HALLTS	Hailing Acoustic Laser and Light Tactical System
IFC	intermediate force capability
JIFCO	Joint Intermediate Force Capabilities Office
LIPE	Laser-Induced Plasma Effects
LRAD	long-range acoustic device
LROI	long-range ocular interrupter
MVSOT	Maritime Vessel Stopping Occlusion Technologies
NATO	North Atlantic Treaty Organization
NCO	non-commissioned officer
NLW	non-lethal weapon
OI	ocular interrupter
PEVS	Pre-Emplaced Electric Vehicle Stopper
RFVS	Radio Frequency Vehicle Stopper
SA	situation awareness
SNS–RDD	Single Net Solution with Remote Deployment Device
UAV	uncrewed aerial vehicle
UGV	uncrewed ground vehicle
VIPER	Vessel Incapacitating Power Effect Radiation
WRENCH	Workbench for refining Rules of Engagement against Crowd Hostiles

References

Acker, Daniella, and Nigel W. Duck, "Cross-Cultural Overconfidence and Biased Self-Attribution," *Journal of Socio-Economics*, Vol. 37, No. 5, October 2008.

Army Nonlethal Scalable Effects Center, "Army Nonlethal Weapons Advanced Planning Briefing to Industry," U.S. Army, September 21, 2017. As of January 10, 2024: https://jnlwp.defense.gov/Portals/50/Documents/Resources/Presentations/ Joint_Integration_Program_Advanced_Planning_Briefs_to_Industry/ 2017%20DoD_OGA%20NLW%20APBI%20Army.pdf?ver=2020-08-25-125259-783

Aros, Susan, Anne Marie Baylouny, Deborah E. Gibbons, and Mary McDonald, "Toward Better Management of Potentially Hostile Crowds," paper presented at the 2021 Winter Simulation Conference, 2021.

Aros, Susan, and Mary McDonald, "Simulating Civil Security Activities in Stability Operations," paper presented at the Interservice/Industry Training, Simulation, and Education Conference, November 2023.

Aruguete, Mara, "Non-Lethal Weapon Use in Crowds," in Lincoln University and the U.S. Army Research Laboratory, *Multi-Task Project to Provide Research Support for Human Research and Engineering Goals Identified by the Army*, 2009.

Axe, David, "Hit the First Vehicle, Hit the Last & Trap the Rest: The Ukrainians Used a Classic Tactic to Devastate a Russian Ammo Convoy," *Forbes*, November 14, 2023.

Beck, Jamal, "New Vehicle Stopper Trials Underway at Tinker Air Force Base," press release, Joint Non-Lethal Weapons Directorate, August 15, 2018.

Bowden, Mark, "The Legacy of Black Hawk Down," *Smithsonian Magazine*, January 2019.

Calder, Simon, "Five Years on from Gatwick Drone Disruption, What Happened—and Could It Happen Again?" *The Independent*, December 19, 2023.

Cialdini, Robert B., *Influence: The Psychology of Persuasion*, 3rd ed., Quill, 1993.

Connable, Ben, Michael J. McNerney, William Marcellino, Aaron B. Frank, Henry Hargrove, Marek N. Posard, S. Rebecca Zimmerman, Natasha Lander, Jasen J. Castillo, and James Sladden, *Will to Fight: Analyzing, Modeling, and Simulating the Will to Fight of Military Units*, RAND Corporation, RR-2341-A, 2018. As of January 10, 2024: https://www.rand.org/pubs/research_reports/RR2341.html

Convention on the Prohibition of the Development, Production, Stockpiling and Use of Chemical Weapons and on Their Destruction, signed at Geneva, Switzerland, September 3, 1992.

Department of Defense Directive 3000.03E, *DoD Executive Agent for Non-Lethal Weapons (NLW) and NLW Policy*, U.S. Department of Defense, incorporating change 2, August 31, 2018.

Dobias, Peter, and Kyle Christensen, "Intermediate Force Capabilities: Countering Adversaries Across the Competition Continuum," *Journal of Advanced Military Studies*, Vol. 14, No. 1, November 2023.

DoD—*See* U.S. Department of Defense.

Drury, John, and Steve Reicher, "The Intergroup Dynamics of Collective Empowerment: Substantiating the Social Identity Model of Crowd Behavior," *Group Processes & Intergroup Relations*, Vol. 2, No. 4, October 1999.

Drury, John, and Steve Reicher, "Collective Action and Psychological Change: The Emergence of New Social Identities," *British Journal of Social Psychology*, Vol. 39, No. 4, December 2000.

Dunning, David, "The Dunning–Kruger Effect: On Being Ignorant of One's Own Ignorance," in James M. Olson and Mark P. Zanna, eds., *Advances in Experimental Social Psychology*, Vol. 44, Elsevier Academic Press, 2011.

Endsley, Mica R., "Toward a Theory of Situation Awareness in Dynamic Systems," *Human Factors: The Journal of the Human Factors and Ergonomics Society*, Vol. 37, No. 1, March 1995.

Executive Order 11850, "Renunciation of Certain Uses in War of Chemical Herbicides and Riot Control Agents," Executive Office of the President, April 8, 1975.

Fairclough, Gordon, "India-China Border Standoff: High in the Mountains, Thousands of Troops Go Toe-to-Toe," *Wall Street Journal*, October 30, 2014.

Filkins, Dexter, and James Glanz, "With Airpower and Armor, Troops Enter Rebel-Held City," *New York Times*, November 8, 2004.

Gain, Nathan, "US Navy Lab Investigates Innovative Non-Lethal Boat Stopping Technology," Naval News, November 25, 2019.

Granovetter, Mark, "Threshold Models of Collective Behavior," *American Journal of Sociology*, Vol. 83, No. 6, May 1978.

Granovetter, Mark, and Roland Soong, "Threshold Models of Diffusion and Collective Behavior," *Journal of Mathematical Sociology*, Vol. 9, No. 3, 1983.

Hirsch, Lily E., *Music in American Crime Prevention and Punishment*, University of Michigan Press, 2012.

Ignatius, David, "Vast Gaza Tunnels Present a Battlefield of Nightmares," *Washington Post*, October 17, 2023.

JIFCO, DoD, Non-Lethal Weapons Program—*See* Joint Intermediate Force Capabilities Office, U.S. Department of Defense, Non-Lethal Weapons Program.

Joint Intermediate Force Capabilities Office, U.S. Department of Defense, Non-Lethal Weapons Program, "Active Denial System FAQs," webpage, undated-a. As of January 10, 2023: https://jnlwp.defense.gov/About/Frequently-Asked-Questions/Active-Denial-System-FAQs/

Joint Intermediate Force Capabilities Office, U.S. Department of Defense, Non-Lethal Weapons Program, "Human Electro-Muscular Incapacitation FAQs," webpage, undated-b. As of January 9, 2024: https://jnlwp.defense.gov/About/Frequently-Asked-Questions/ Human-Electro-Muscular-Incapacitation-FAQs/

Joint Intermediate Force Capabilities Office, U.S. Department of Defense, Non-Lethal Weapons Program, "Oleoresin Capsicum Dispensers," webpage, undated-c. As of January 9, 2024: https://jnlwp.defense.gov/Current-Intermediate-Force-Capabilities/ Oleoresin-Capsicum-Dispensers/

Joint Intermediate Force Capabilities Office, U.S. Department of Defense, Non-Lethal Weapons Program, "Variable Kinetic System (VKS) Non-Lethal Launcher System," webpage, undated-d. As of January 9, 2024: https://jnlwp.defense.gov/Current-Intermediate-Force-Capabilities/Variable-Kinetic-System/

Joint Intermediate Force Capabilities Office, U.S. Department of Defense, Non-Lethal Weapons Program, *DoD Non-Lethal Capabilities: Enhancing Readiness for Crisis Response: Annual Review*, 2015.

Joint Intermediate Force Capabilities Office, U.S. Department of Defense, Non-Lethal Weapons Program, "Non-Lethal Optical Distractors Fact Sheet," May 2016.

Joint Intermediate Force Capabilities Office, U.S. Department of Defense, Non-Lethal Weapons Program, "Acoustic Hailing Devices Fact Sheet," November 16, 2018a.

Joint Intermediate Force Capabilities Office, U.S. Department of Defense, Non-Lethal Weapons Program, "Radio Frequency Vehicle Stopper," November 16, 2018b.

Joint Intermediate Force Capabilities Office, U.S. Department of Defense, Non-Lethal Weapons Program, "Single Net Solution with Remote Deployment Device," November 16, 2018c.

Joint Intermediate Force Capabilities Office, U.S. Department of Defense, Non-Lethal Weapons Program, "Vessel-Stopping Prototype," November 16, 2018d.

Kahneman, Daniel, *Thinking, Fast and Slow*, Farrar, Straus and Giroux, 2013.

Kahneman, Daniel, Paul Slovic, and Amos Tversky, eds., *Judgment Under Uncertainty: Heuristics and Biases*, Cambridge University Press, 1982.

Krechetnikov, Artem, "Moscow Theatre Siege: Questions Remain Unanswered," BBC, October 24, 2012.

Kruger, Justin, and David Dunning, "Unskilled and Unaware of It: How Difficulties in Recognizing One's Own Incompetence Lead to Inflated Self-Assessments," *Journal of Personality and Social Psychology*, Vol. 77, No. 6, 1999.

LeVine, Susan, *The Active Denial System: A Revolutionary, Non-Lethal Weapon for Today's Battlefield*, National Defense University Center for Technology and National Security Policy, June 2009.

Lewin, Kurt, *Resolving Social Conflicts and Field Theory in Social Science*, American Psychological Association, 1997.

Lin, Bonny, Cristina L. Garafola, Bruce McClintock, Jonah Blank, Jeffrey W. Hornung, Karen Schwindt, Jennifer D. P. Moroney, Paul Orner, Dennis Borrman, Sarah W. Denton, and Jason Chambers, *Competition in the Gray Zone: Countering China's Coercion Against U.S. Allies and Partners in the Indo-Pacific*, RAND Corporation, RR-A594-1, 2022. As of January 10, 2024: https://www.rand.org/pubs/research_reports/RRA594-1.html

Mapp, Katherine, "Promising New Tool Protects Ships, Sailors," Naval Sea Systems Command, November 21, 2019.

Martinez, Luis, "Chinese Warship Cuts Off US Navy Ship, Marking 2nd Military Provocation in Week," ABC News, June 4, 2023.

Mattis, James, *Summary of the 2018 National Defense Strategy: Sharpening the American Military's Competitive Edge*, U.S. Department of Defense, 2018.

Mazarr, Michael J., Joe Cheravitch, Jeffrey W. Hornung, and Stephanie Pezard, *What Deters and Why: Applying a Framework to Assess Deterrence of Gray Zone Aggression*, RAND Corporation, RR-3142-A, 2021. As of January 10, 2024: https://www.rand.org/pubs/research_reports/RR3142.html

Mehta, Aaron, "Two US Airmen Injured by Chinese Lasers in Djibouti, DoD Says," Defense News, May 3, 2018.

Mezzacappa, Elizabeth S., *Crowd Dynamics and Military Interaction with Non-Lethal Weapons and Systems*, Target Behavioral Response Laboratory, JNLWD11-006, 2009.

Monaghan, Sean, *Deterring Hybrid Threats: Towards a Fifth Wave of Deterrence Theory and Practice*, European Centre of Excellence for Countering Hybrid Threats, Hybrid CoE Paper 12, March 2022.

Moore, Don A., Amelia S. Dev, and Ekaterina Y. Goncharova, "Overconfidence Across Cultures," *Collabra: Psychology*, Vol. 4, No. 1, October 2018.

Morris, Lyle J., Michael J. Mazarr, Jeffrey W. Hornung, Stephanie Pezard, Anika Binnendijk, and Marta Kepe, *Gaining Competitive Advantage in the Gray Zone: Response Options for Coercive Aggression Below the Threshold of Major War*, RAND Corporation, RR-2942-OSD, 2019. As of January 10, 2024:
https://www.rand.org/pubs/research_reports/RR2942.html

Myre, Greg, "How the U.S. Military Used Guns N' Roses to Make a Dictator Give Up," NPR, May 30, 2017.

Office of General Counsel, U.S. Department of Defense, *Department of Defense Law of War Manual,* updated May 2016.

Plous, Scott, *The Psychology of Judgment and Decision Making*, McGraw-Hill, 1993.

Protocol on Blinding Laser Weapons (Protocol IV to the 1980 Convention), signed October 13, 1995.

Romita Grocholski, Krista, Scott Savitz, Nancy Huerta, Alyson Youngblood, Jonathan P. Wong, Sydney Litterer, Raza Khan, and Monika Cooper, *Logic Model for Non-Lethal Weapons in the U.S. Department of Defense*, RAND Corporation, DV-A1544-1, 2023. As of January 10, 2024:
https://www.rand.org/pubs/visualizations/DVA1544-1.html

Romita Grocholski, Krista, Scott Savitz, Sydney Litterer, Monika Cooper, Clay McKinney, and Andrew Ziebell, *Assessing the Impact of Diverse Intermediate Force Capabilities and Integrating Them into Wargames for the U.S. Department of Defense and NATO*, RAND Corporation, RR-A1544-1, 2023. As of January 10, 2024:
https://www.rand.org/pubs/research_reports/RRA1544-1.html

Romita Grocholski, Krista, Scott Savitz, Jonathan P. Wong, Sydney Litterer, Raza Khan, and Monika Cooper, *How to Effectively Assess the Impact of Non-Lethal Weapons as Intermediate Force Capabilities*, RAND Corporation, RR-A654-1, 2022. As of January 10, 2024:
https://www.rand.org/pubs/research_reports/RRA654-1.html

Savitz, Scott, Miriam Matthews, and Sarah Weilant, *Assessing Impact to Inform Decisions: A Toolkit on Measures for Policymakers*, RAND Corporation, TL-263-OSD, 2017. As of January 10, 2024:
https://www.rand.org/pubs/tools/TL263.html

Schelling, Thomas, *The Strategy of Conflict*, Harvard University Press, 1960.

Shelbourne, Mallory, "U.S. Navy Sends Nontraditional Ships to Support Sudan Evacuation," U.S. Naval Institute News, May 1, 2023.

Silver, Michael, *Tactics, Training, and Procedures for the Warfighter Reacting to Crowd Dynamics*, Anacapa Sciences, Inc., 2002.

Simons, Daniel J., and Christopher F. Chabris, "Gorillas in Our Midst: Sustained Inattentional Blindness for Dynamic Events," *Perception*, Vol. 28, No. 9, September 1999.

Stott, Clifford, Otto Adang, Andrew Livingstone, and Martina Schreiber, "Tackling Football Hooliganism: A Quantitative Study of Public Order, Policing and Crowd Psychology," *Psychology, Public Policy, and Law*, Vol. 14, No. 2, May 2008.

Stott, Clifford, and Stephen Reicher, "Crowd Actions as Intergroup Process: Introducing the Police Perspective," *European Journal of Social Psychology*, Vol. 28, No. 4, December 1998.

"Syria War: Raqqa Deal Agreed to Evacuate Civilians," BBC, October 14, 2017.

Teigen, Karl Halvor, "Yerkes-Dodson: A Law for All Seasons," *Theory & Psychology*, Vol. 4, No. 4, November 1994.

Tucker, Patrick, "The US Military Is Making Lasers That Create Voices out of Thin Air," Defense One, March 20, 2018.

Tversky, Amos, and Daniel Kahneman, "Availability: A Heuristic for Judging Frequency and Probability," *Cognitive Psychology*, Vol. 5, No. 2, September 1973.

Tversky, Amos, and Daniel Kahneman, "The Framing of Decisions and the Psychology of Choice," *Science*, Vol. 211, No. 4481, January 1981.

Tversky, Amos, and Daniel Kahneman, *Choices, Values, and Frames*, Cambridge University Press, 2000.

U.S. Department of Defense, *DOD Dictionary of Military and Associated Terms*, Joint Chiefs of Staff, March 2017.

U.S. Department of Defense, Physical Security Enterprise and Analysis Group, "Hailing Acoustic Laser & Light Tactical System (HALLTS)," webpage, undated. As of January 5, 2024: https://www.acq.osd.mil/ncbdp/nm/pseag/Hailing%20Acoustic%20Laser%20&%20Light%20Tactical%20System%20(HALLTS).html

Voice of America, "US Confirms Egyptian Death in Suez Canal Shooting," November 1, 2009.

Voice of America, "Kerry: Syrian Gas Attack Crossed 'Global Red Line,'" September 8, 2013.

Wahn, Basil, and Peter König, "Is Attentional Resource Allocation Across Sensory Modalities Task-Dependent?" *Advances in Cognitive Psychology*, Vol. 13, No. 1, March 2017.

Werner, Ben, "Destroyer USS Decatur Has Close Encounter with Chinese Warship," U.S. Naval Institute News, October 1, 2018.

Wickens, Christopher D., *Engineering Psychology and Human Performance*, 2nd ed., HarperCollins Publishers, 1992.

Yeung, Jessie, "Indian and Chinese Troops Fight with Sticks and Bricks in Video," CNN, December 15, 2022.

Zargar, Arshad R., "India-China Border Standoff Turns Deadly for First Time in Decades," CBS News, June 16, 2020.